*Life ∾ The
Unfinished
Experiment*

*Au fond de l'Inconnu pour trouver du nouveau*
—Charles Baudelaire, *"Le Voyage"*

(Into the depths of the Unknown in quest of something new
—translation by Francis Scarfe)

# Life ∽∽∽ The Unfinished Experiment

## S. E. LURIA

*CHARLES SCRIBNER'S SONS*
*NEW YORK*

*To Zella, a unique life partner*

# Contents

*Life ෴ The
Unfinished
Experiment*

# Introduction

This book owes its origin to a suggestion by Theodosius Dobzhansky that he and I collaborate in writing a pair of books on
"Life and Man." The idea was to present to the intelligent layman
the concepts of a science that is progressing so rapidly as to make
its findings increasingly relevant to human society. Coming from
a great geneticist who is also a distinguished writer, the suggestion
appealed to me and I accepted it. I had drafted the first two
chapters when *Chance and Necessity* by my friend Jacques Monod
appeared, a book that quickly and somewhat surprisingly became
an overnight best-seller in France and was almost a best-seller in
the United States as well. I say surprisingly, because *Chance and
Necessity,* though an important book, is not an easy one to read.
It deals with the central problem of biology—the nature and
function of the hereditary material and its relation to evolution—
in philosophically stimulating terms but with hardly any conces-
sion to the non-scientist. If people could grasp the science con-

densed in that book, was there any need, I asked myself, for a book on life such as I had undertaken to write?

Inquiries made both in the United States and in France convinced me of a different truth: most people who had bought Monod's book had not really read it. At least, they had not read the technical, biochemical sections in the book. They had agreed or disagreed vehemently with points of view presented in the philosophical chapters and accordingly had put the book on prominent display on a coffee table or into the semi-oblivion of a bookshelf.

The present book—unfortunately, Dobzhansky has as yet been unable to write the companion book, which was to deal with organismic aspects of life—would have made the task easier for the readers of *Chance and Necessity*. It tries to present a simple, factual view of modern biology as interpreted in terms of molecular mechanisms—the so-called molecular biology—and dealing with the workings of life from the gene to the cell to the complex organism and to the species. The facts of molecular biology provide man an understanding of the historical process of life of which he is part, of the functioning of his own body, of his society, of the living environment with which he must perpetually come to terms. Ultimately they will reveal even the secrets of the functioning of the human mind.

Inevitably this book makes frequent reference to that remarkable species, *Homo sapiens*. Evolution has brought about, in this as in most other species, sexual reproduction accompanied by sexual dimorphism. In deference to the well-justified concern for equality between sexes even in the relatively trivial matter of the use of words, I tried to avoid the use of "man" and "his" as referents to the human species as a whole but found no way to do so without being bogged down in the clumsiness of alternative expressions. So "man" it is, in the deplorable sexist tradition of the

English language, a tradition that proves hard to overcome. "Man" stands for the species—that is, about equal numbers of women and men. An understanding of biological facts and an awareness of the depth of biological interdependence among all members of a species may help foster the recognition of the justness of social equality between members of the two sexes and among all groups of mankind.

Science's present grasp of the molecular mechanisms of life, however incomplete, represents a remarkable intellectual achievement, a satisfyingly coherent set of hypotheses and explanations. Moreover, it is intensely relevant—as a source of wisdom as well as of dangers—to a range of problems that already beset modern man or will probably do so in years to come: from the manipulation of his own heredity, to the possible control of his demographic expansion, to the effective learning to live in a balanced environment.

Life has two scientific aspects: life in action and life in time. Life in action is the functioning of living organisms, the molecular and atomic events brought about by the presence of life, and is the subject matter of biochemistry. Life in time is the persistence and disappearance and replacement of organisms, by individual death as well as by the generation and differential proliferation of new species—in one word, evolution. These two aspects, biochemistry and evolution, make life a unique phenomenon in the history of the earth, one that long before the coming of man had impressed its profound mark on the features, the climate, the very structure of the planet earth.

Life is distinct from all other natural phenomena in one feature: it has a program. All other natural phenomena occur more or less at random, like the movement of clouds in the changing winds, or the disintegration of radioactive atoms, or the collision of molecules in a heated fluid. When physical phenomena have a regular trend, it is generally a trend toward increased disorder, in conformity with the physical law that affirms the tendency toward a minimum of molecular order. Even when order appears to increase, as in the crystallization of substances from solutions, it is a repetitious, monotonous, noncreative order. Nonbiological phenomena are characteristically the expression of the statistical behavior of large numbers of units, not of the unique performance of a single object or structure.

Only in life does individuality emerge. Life's program unfolds itself in the wonderful growth of a germ into an organism, in the blossoming of a species to fill an environmental setting, in the creative replacement of species with species in the course of evolution—because life's program has been inscribed in a unique substance, the substance of the genes. This material substrate of life is an exception, not to the laws of physics and chemistry but to the run-of-the-mill types of molecules. It is a substance whose construction insures both stability and almost infinite variety of individual patterns. It provides for copying with an accuracy unattainable in any other known molecular species. At the same time, this program substance is capable of change, and its changes become the basis for biological evolution.

The program of life, embodied in the substance of the genes and expressed in the forms of organisms and in their evolution, is not like the conscious programs of human enterprises, be they works of art or social undertakings. It is not a project for the future: it is an inventory of the present, an array of potentialities embodied in the substance of the genes. As stated by the French geneticist François Jacob in *La Logique du Vivant* (1970), "the individual becomes the realization of a program prescribed by heredity." For a single organism, the program is the inborn plan that controls its development and its life functions, modifiable of course by external influences. For a species, the program is the entire range of genetic types it encompasses, a range that determines whether that species, in given surroundings, will persist or flourish or die out.

The dualism of the material nature of life's program on the one hand and the historical nature of biological evolution on the other hand is the leit-motif of this book. Life evolved, reached its present state, and will further develop by the creative interplay of the chemical workings of the genetic material and the historical workings of the natural forces that favor now one species, now another, promoting any biochemical invention that provides increased fitness. Stupendous devices such as the brain and mind of man are biochemical inventions as challenging and as mysterious as those that produced the equally stupendous social organization of insects. To the scientist, the uniqueness of man is purely a biological uniqueness rather than the superposition of something nonbiological—soul or spiritual essence—upon the workings of biological evolution. The nature of the mechanisms responsible for these highly complex phenomena still escapes the biologist, but he is confident that this will not always be so.

The science of heredity is less than one hundred years old, modern biochemistry less than fifty, molecular biology barely twenty, and their progress has been astonishingly rapid. As one looks back at the millennia of ignorance and forward toward the harvest of knowledge still to be gained, both pride and humility are in order. As he gains understanding of life and of himself, man seems to be well on his way to fulfill the prediction of Genesis: "Ye shall be as gods, knowing good and evil."

But man's knowledge of himself is still scanty and is blurred by the fogs of legend and superstition—naive but inevitable attempts of his ancestors to acquire knowledge by intuition rather than by reason. Meanwhile the course of events proceeds swiftly. Man may soon be called upon to make choices on at least some aspects of his own biological future, realizing that the old intuitions do not suffice and that scientific knowledge, partial as it may be, is the truly reliable tool at his disposal. One motive for the writing of this book is the belief that scientists have a responsibility to inform the public of the state of their knowledge, especially when that knowledge becomes relevant to the welfare of mankind.

# 1 ✍ Evolution

In almost any New England village you are likely to find an antique shop (if you are near a main highway, "ye antique shoppe"). If the shop is a good one, you may buy something from the past—a very recent past, compared to that represented by Angkor Wat, or Stonehenge, or the sculptures in the Assyrian rooms in the British Museum. Yet for the educated person all past has an appeal that is as hard to escape as it is to explain. An object, irrespective of its beauty, has meaning because of its age—because of having survived to give tangible proof that the past was real and durable.

An antique object is considered even more valuable if it is rare or unique. Most things out of the past disappear; the few that remain become the evidence of archaeology or history. Through them people today can imagine the life and habits and emotions of their forefathers. Viewing and fondling the artifacts of other ages—a family souvenir, a piece of furniture two centuries old, a clay pot from Etruria, or an obsidian arrowhead from Mexico— evokes hidden resonances from personal experiences. Traces of that past history exist in modern men and women—in their thoughts and emotions as well as in their bodies. They thus become aware of being part of a continuum, extending back in time and forward into the future, a continuum of which those now living are so to speak the cross section at the present time.

In fact, like the artifacts of the past that survive to become antiques, the human beings and other organisms that are alive today represent an exceptional sample from a very large range of possibilities. It will help to understand life, and evolution, which is the central feature of life, to reflect about history, not as the complete record of past events, but as the incomplete record of the range of possible events that might have occurred. This is the chief distinction between classical academic history and the work of modern historians, who inquire why certain events occurred and others did not, or what economic forces, what personalities, what chance happenings shaped the unique course of human affairs, selecting out of the myriad of potential events the unique set of those that actually came true.

Even more relevant to learning to look at life as a historical process is understanding another quality of such processes: their irreversibility. Each present, each cross section of history in time, is the unique outcome of the actual events of the past and at the same time the bottleneck of all the future. Unexploited opportunities may present themselves again, but if they do it will be in a new context, under new circumstances, to new generations whose life and culture will be the products of those experiences that have actually been lived and that have left their imprint on the present and on the future.

Like human history, life also is a historical process. The

living organisms of today are the incomplete record of the possibilities of the past. The smallest bacteria, the humblest worms
and snails, algae and mosses, as well as the proudest trees, the
most gorgeous birds, and the billions of human beings are a sparse
sample of the total range of living things that might have existed.
Individual men and women often experience a strange nostalgia
—a thought of what might have been, a longing for past opportunities either missed or never available, and even more often a
longing for horizons that might have opened up, if only . . .

And yet, how many ever stop to think how remarkable it is
that they should exist at all? Each human being is the actualizaton
of an extremely improbable chance—in fact, of a series of improbable chances, extending all the way back to the unique event that
more than three billion years ago started life on earth on its chancy
course.

Among certain orthodox Jewish groups, I am told, there used
to be a custom that sons should not follow their father's bier. Some
early rabbi apparently taught that the funeral procession of each
man is watched by the souls of those children of his that might
have been and never were—seed spilled or unused—and these
should not be given the chance to curse their living brothers with
their envy: "Why they and not we?"

There is biological wisdom in this odd superstition: each
man, each organism, that dies without living descent is a biological
dead end. Among modern men in prosperous societies and also
among valuable domestic animals, a majority of the newborn live
and have progeny in their turn. Biologically, however, this is an
exceptional situation. Throughout the range of living things, from
plants to animals, all the way to the poor people in the jungles of
South America, the rice paddies of Asia, and the urban slums of
the rich countries, most of what is born dies in youth without

contributing its biological heredity to future generations. Biological evolution is the sequence of the few successes that emerge from the mass of reproductive failures.

The idea of biological evolution, which was placed on a solid scientific basis by Charles Darwin with the publication of *On the Origin of Species* in 1859, was not new. Yet it was not the kind of idea that needs only to be formulated to impose itself by its obvious truth. Resistance to it came not only from fundamentalist Bible worshippers, but also from human conceit, since evolution threatened to relegate men to an animal species and offended them by proclaiming their relationship to apes. Moreover, the idea was resisted by something deeper in the structure of human culture, except among those enlightened elements of society which, following the Renaissance and the eighteenth-century rationalist movement, had integrated into their world view the concept of a directional historic process. The idea of history as progression rather than as static repetition was still basically alien to the mainstream of human culture. Greco-Roman thought never saw the passage of man in time in fully historical terms. Nor was the idea of history as an irreversible developmental process part of the conception of the great Eastern cultures. For these, as for the Greeks, history remained a series of episodes illustrating the struggles of an immutable mankind against an inexorable fate, each man in a universe unto himself, at best struggling, as in the Buddhist view, to achieve an ultimate state of individual perfection by merging with an insensitive cosmos. The aggressive entry onto the world scene of Jewish theology and its Hellenistic offspring, Christianity, did inject a historical element into the world view of Western civilization. Creation, the original sin, redemption, and the final judgment provided a directional arrow to man's history. Yet Christian thought remained fundamentally unhistorical. Each man was on earth by the will of God to fulfill an individual test of good and evil. To bring the Kingdom of God to earth required the miraculous incarnation of God, the Son. Change, evolution, and even technological progress were certainly there for everyone to see and make use of in the business of day-by-day living, but they were not part of the dominant philosophies. The great achievements of mankind—metal working, clay firing, and above all agriculture and the domestication of animals—dated from prehistory and in the Judeo-Christian myth were subsumed into the epic of creation. In six days the stage was set, and by the time of the death of the first man all the basic arts had been practiced and the world had been set onto its unchanging course.

It is not surprising that human thought followed such a path. Because of its leisurely time scale, organic evolution is not an obvious process. Rather, the model for a world view became the cosmology of the heavens, the mechanical motion of the planets in the immutable panorama of the firmaments—immutable except for rare anomalies, supernovae or long-period comets, whose appearance was believed to signal sporadic willful interventions of divinity into the course of human affairs.

The concept of history as a directional process entered the realm of human affairs in the early seventeenth century, mainly with the writings of the Italian scholar Giambattista Vico (1668–1744). His theory of historical cycles was not a restatement of a Greek view of monotonously repeating cycles in the course of events, but an analytical approach to the study of human affairs. Vico's was the first view of history fashioned to fit the heliocentric cosmology of Copernicus, Galileo, and Newton; the humanistic philosophies of Descartes and Spinoza; the dynamic, expansionist society of explorers, conquerors, and rising manufacturers, and the related activist view of personal responsibility embodied in the

Protestant religion. The historical thinking of Vico and his successors, permeating the whole intellectual life of the seventeenth and eighteenth centuries, evolved into the Utopian visions of the philosophers of the French Enlightenment and ultimately into the stirring perspectives of nineteenth-century Marxism. It established man as the protagonist of history seen as a developmental process, replacing man as an individual soul testing his eligibility to salvation in his all-important passage on earth.

Yet, before Darwin, these ideas failed to penetrate either the emotional consciousness of the man in the street or the textbooks of the theologians charged with the education of the young. The fact that in the new cosmology man stood revealed as the rational inhabitant of a minor satellite of one star among myriads could readily, if irrationally, be seen as a marvelous example of the mysterious ways of the Creator. But the concept of human history as a developmental process was much more dangerous. If mankind collectively, rather than man as an individual, was part of a process, the question of origins became as pressing as that of destiny. The goal might be perfection, but what was the source? If it was Creation, was the product an imperfect child of God, launched into a perilous course to seek ultimate perfection on earth rather than salvation in heaven?

Upon the thinking of the nineteenth century, the theory of biological evolution burst as the logical but unwelcome culmination of the historical view of the world. It explained that not only human society but the whole world of living things had a history, which was not determined by some purpose or direction in the future, but only by events of the past. It allowed no reason for existence other than having been the successful product favored by natural selection. Darwin's great insight, the recognition of natural selection as the determinant of what exists as a purely statistical force blindly directing the everchanging panorama of the living world, was not easy to accept. It made the present, not the gateway to a hopeful future, but the chancy outcome of past escapes. By integrating man into his comprehensive record of biological evolution Darwin dispelled the hope of some immanent purpose in human history. However unique man's endowment with consciousness might make him, his past and future represented only the passing of one species on the surface of the earth. They had no reason, no goal, no meaning, except what man in his existential freedom might choose to set for himself.

The ground was ready for the theory of evolution, but it was a ground strewn with obstacles. Evolution explains the present as the outcome of the past; what is, in terms of what was. It explains but it makes no promises. The most challenging and resistance-provoking thought is not the identification of man with the animal world or his relationship with the apes, but the substitution of reasons for survival in place of reasons for having been born; of causes in place of purpose; of an all-determining past in place of an all-important future.

The historical view of man's destiny that had developed in the Enlightenment was not truly evolutionary. It substituted history for heaven-decided purpose as the determinant of the course of human events. History as a directional process, mechanistically oriented toward some higher good of mankind, tended to replace God as the source and justification of values for human activities. As the modern French writer Albert Camus has pointed out in *The Rebel,* the metaphysics of history simply replaced the metaphysics of religion in Western man's search for an ultimate good.

In biology, the theory of evolution was a great unifying generalization. It brought together the whole living world, past, present and future, into one single record of descent—the record of the conquest of almost all possible environments on the globe, of the increasing numbers of living things, of the ever more pervasive presence of the organic at the expense of the inorganic. But, contrary to superficial belief, the record of biological descent is not in the main a record of successes. Rather, it is one of innumerable failures interspersed with relatively few but all-important lucky breaks. As noted earlier, the range of living things present at any one time is the record of the few survivors among many extinguished lines of descent. What exists at any one time is the descendants of those few that proved their fitness at earlier times.

Few concepts lend themselves to greater misunderstandings than does that of biological fitness. Because of the semantic ambiguity, we tend to identify evolutionary fitness with physical fitness such as may be provided by exercise, or with intellectual fitness as cultivated by education. But to the geneticist evolutionary fitness is a narrower, more precise concept: it is a measure of the number of descendants that will represent an individual (or a group of individuals, or a species) in the range of organisms in existence at a specified later time. The best educated and most superbly conditioned individual, if he dies childless, is evolutionarily a failure. And even evolutionary success may be transient, as is shown by the extinction of organisms that were highly successful in earlier times.

Thus fitness can be given a precise value in terms of numbers and times: if $A$ and $B$ are two contemporary members of the same species, $A$ is evolutionarily more fit than $B$, as judged at time $t$, if the number of $A$'s descendants at that time is larger than $B$'s. Natural selection is the emergence and flourishing of certain populations and species that manage by reproductive success to reach large numbers in certain specific environments over reasonably long periods of time. If a population of a given species is reduced, through migration or other accident, to a very small number of individuals, those individuals, even if originally they were not the best-endowed from a reproductive standpoint, may turn out to be the progenitors of a whole new line of descent. This kind of success, resulting from the chance separation of a small group of individuals, has been called *genetic drift*. In itself genetic drift is a perturbing mechanism in evolution. By magnifying the role of chance over that of effective selection in large populations, it limits the precise adaptive molding of species to their environment. It tends to slow down the work of natural selection. In most cases, however, the successful populations are the large ones, and genetic drift has a relatively minor role in the over-all pattern of evolution.

The role of the environment in evolution is crucial and readily evident: a given environment generates fitness to itself by favoring the multiplication of those individuals whose genetic make-up renders them best suited to life in that environment. A population of organisms and its environment, therefore, constitute an interacting system. It is proper to speak of "the fitness of the environment" (the title of a remarkable book written early in the twentieth century by the American biochemist Laurence J. Henderson) in the sense that the environment in which certain organisms are found appears almost to have been manufactured, so to speak, to fit those organisms that are so well adapted within it. Actually, however, the environment is most often the main agent in the adjustment process. The organisms provide the range of biological choices on which the environment acts selectively to bring forth the most successful types. Men (and to a smaller extent some other species of animals, such as beavers) have the ability to mold their environment and make it more suitable to their own goals. But even for organisms that cannot do this, the interaction between environment and organism is always at play. Each new generation bears witness in its genetic make-up to the selecting and perfecting role of this interplay.

The process that generates, within a given species as well as in different species, the range of genetic make-ups among which natural selection works is heritable variation. It is important to bear in mind that not all biological variation is heritable. Suppose I have an identical twin—that is, my twin brother and I were produced from the same fertilized egg. If I were to live all my life in the tropics, my skin color would probably be darker than that of my twin if he were raised in New England. But our children would not inherit my ruddiness or his pallor. Genetically, identical twins remain identical. The inheritance of acquired characteristics, proposed by the great French zoologist Jean-Baptiste de Lamarck (1744–1829) and provisionally if hesitantly accepted by Darwin for lack of a better theory, has been proven untenable. Innumerable experiments have given results completely incompatible with the inheritance of characters acquired during life. What is inherited is a genetic potential, a set of genes or, in technical terms, a *genotype*. Each individual's genotype determines the potential range of his functional adaptations to his environment.

A bacterium or an amoeba, which reproduces by splitting, has two descendants that are genetically identical to the single parent (except for the occurrence of genetic changes, as explained later). In organisms with sexual reproduction, a pair of individuals of different sexes, if they mate successfully, have offspring that derive their genotypes partly from one parent, partly from the other, according to the rules of genetics. If the parents differ in a number of characters—that is, in the structure of certain genes— there is opportunity at each generation for new combinations. Thus sexual reproduction accelerates the rate of evolution because it multiplies the range of genetic variability present in a population.

There are two exceptions to this rule: self-fertilization and mating within inbred lines. Self-fertilization, which occurs mostly in plants, is not a true exception because it is not really a form of sexuality. Matings within an inbred line are interesting because they serve to illustrate the rule.

Inbred lines of animals can be created in the laboratory for purposes of scientific research by repeated brother-to-sister mating in mice, rats, rabbits, and other small animals. After many repeated rounds of such incestuous generations, the animals of any one line of descent become more and more similar genetically because their "gene pool" becomes progressively more restricted. The number of generations needed to get fewer than any desired percentage of gene differences in a group of animals can be calculated, the percentage becoming smaller as the number of brother-to-sister matings increases. Finally the point is reached at which, for all practical purposes, all animals in a colony, apart from sex differences, are as similar to each other genetically as identical twins. The inbred line has become practically a pure line of descent. Mating among such animals and their descendants, however pleasurable it may still be to them, ceases to fulfill one evolutionary purpose of sex. It becomes, genetically speaking, like the splitting of an amoeba, because two identical parents must generate identical progeny.

The differences that exist in nature among individuals and among species are the results of changes, or *mutations,* in the genetic material. Mutations occur all the time but at very low frequency, so that a given gene may undergo a mutation only once in several thousand generations. These mutations, which result from the intrinsic vulnerability of the chemical structure of the genetic material as well as from errors in the copying of this material, are the source of all genetic variability, which in turn provides the material substrate of evolution by natural selection.

To return to evolution, its most remarkable feature is its apparent precision—that is, the almost uncannily precise adaptation it generates. Each living organism appears to have been manufactured for exquisitely adjusted performance in its natural environment.

This feat is accomplished by the operation of natural selection through the so-called law of large numbers. The process of genetic mutation is strictly random; the reshuffling of gene groups at each mating (except for matings within pure lines) is largely if not completely random; the occasional spurious fitness accruing to certain individuals in small populations—genetic drift—is due to chance. But the main force, natural selection, is anything but random. Choosing among the enormous numbers of gene combinations present in any large population, natural selection, by its continuous unobtrusive performance, causes the population to become progressively more fitted to its environment because, generation after generation, the bearers of the more successful genotypes have relatively more descendants. This is of course a circular argument, since by definition evolutionary fitness is equated with relative abundance of descendants. But the fact is that, in relatively unchanging environments, a species does become progressively more adapted and better specialized. Natural selection does work. Moreover, it continues to work even when the environment changes, because the lottery of the sexual-mating process preserves, within every natural population, a reservoir of heritable variation that provides a range of genotypes. Only major changes of environment, such as glaciations or the appearance of new predators or of new infectious diseases, can bring populations, species, and even entire genera and families of organisms to an evolutionary end. This happened at the end of the Cretaceous period to the dinosaurs that roamed meadows, sky, and seas about 100 million years ago. Less dramatically, it happened to the great majority of lines of descent, including man's humanoid ancestors and all the branches to which they gave rise except for the single one that stayed and flourished, *Homo sapiens*.

In populations of asexual organisms, which include most bacteria, there is no reshuffling of genetic characters at each generation, and only the existence of very large numbers of individuals, including fair numbers of mutant types, makes possible an effective adaptation to new environments. In general, the key to the working of natural selection, to the fact that it generates adaptation rather than chaos, is the law of large numbers. Large populations, by including a range of genotypes, provide the opportunity for the effective work of differential reproduction.

This is a deceptively simple but actually rather subtle concept, which must be grasped clearly in order to understand life, including man's. Not all combinations of genes need to be present in large numbers in order for selection to work effectively; in fact, any natural population of sexually breeding organisms contains no two exactly duplicate genotypes (except for identical twins). What selection accomplishes is to increase, in successive generations, the frequency of those genes which, in a greater variety of combinations with other genes, tend to foster a more fruitful reproductive career. If a given gene increases the reproductive success of the individuals that carry it, it will be present in larger numbers in successive generation. Selection does not act directly on the genes; it works on *phenotypes,* on the complex of the actual characteristics of the organism. The phenotype of an organism is the expression of its genotype within the environmental and developmental history of that organism. It is not a structure or a sum of structures; it is a pattern of organization. In fact, what evolves is not sets of structures but patterns of organized form and function. Yet selection of reproductively successful phenotypes brings about an enrichment, within the population as a whole, of certain genes that increase fitness, as defined here, in a great number of combinations with other genes.

Selection acts blindly but effectively. Seen in retrospect, its workings manifest uncanny precision. But this precision is like that with which one can calculate the chance of having three of a kind in a poker hand. Only because of large numbers can the likely outcome becomes realities, just as only during an extremely long card game will the various combinations of cards come up at nearly their expected frequencies. The probabilistic character of the game, whether poker or biological survival, is converted to a quasi-deterministic quality by the large number of chances. This is true of the past as well as of the future. Many people, including some scientists, have refused to believe that a probabilistic process like natural selection could have worked with such precision to bring about the almost uncanny fitness of plants and animals to their natural environments, as well as the marvels of the human mind. They have suggested the possible existence of biological laws other than those of physics and chemistry in order to explain the direction, speed, and apparent purposefulness of evolution. But what seems to be purposefulness is only selection for superiority of performance, however slight, that leads to improved reproductive success. Invoking unknown biological laws to explain the efficiency of natural selection is a return to vitalism, the theory that tried to explain the uniqueness of living organisms by postulating a "vital force." Such explanations explain nothing, and, in the ultimate analysis, they can be traced to the metaphysical belief that each organism has a vital spirit or soul imposed upon it from outside.

The modern theory of evolution, like all historical theories, is explanatory rather than predictive. To miss this point is a mistake that theoreticians of history have often made. Prediction in evolution would require not only a knowledge of the main force —natural selection—but also a prescience of all future environmental conditions, as well as of future balances between the quasi-deterministic effects of the law of great numbers and the purely probabilistic role of genetic drift.

For evolution, like history, is not like coin tossing or a game of cards. It has another essential characteristic: irreversibility. All that will be is the descendant of what is, just as what is comes from what has been, not from what might have been. Men are the children of reality, not of hypothetical situations, and the evolutionary reality—the range of the organisms that actually exist— is but a small sample of all past opportunities. If a species dies out, evolution may, in an environment where the lost species would have been fitted, fashion a reasonable likeness of it. This process, named *convergence,* is the one that gave to dolphins and whales their almost fishlike shape and to bats their birdlike appearance. But evolution does not thereby retrace its steps. It makes the best of whatever genetic materials have managed to remain available to it at any one time.

Perhaps the sense of dismay that man, the one conscious animal, feels at dying without offspring and the pride that he feels in his descendants are expressions of an unconscious, internalized feeling of the continuity of evolution through the passing self.

# 2 ∾ *Heredity*

An amoeba grows, divides, and gives rise to two amoebae, each like the original one. If an inheritable change occurs, the changed amoeba generates a new line of descent, which may either die out or persist and evolve further, depending on its fitness to the environment, its ability to compete effectively for nutrients, its resistance to adverse conditions—in a word, its success in the play of natural selection. One might believe that the heredity of the amoeba or of any other one-cell organism that reproduces by simple fission is its entire structure, half of which is given to each of the daughter cells at each generation.

Most organisms, however, reproduce in a different way. In plants and animals, including man, two individuals of opposite sexes contribute *germ cells,* egg and sperm respectively, to generate a new organism. This new organism is not the sum of the heredities of the two parents: it is a combination of parts of the heredity of each. Within the range of forms that individuals of the species *Homo sapiens* can assume, a child will resemble one parent in its sex but in other characteristics may resemble its father in some, its mother in others, and neither of them in still others.

The science of genetics takes as its point of departure the mystery of family resemblances and, using controlled experimental breeding in animals or plants, analyzes the underlying mechanisms. It is a young science, whose foundations were laid by a Bohemian monk, Gregor Mendel (1822–1884), at about the same time that Darwin published *On the Origin of Species.* But whereas Darwin's book hit the scientific and even the general community like a bombshell, Mendel's work remained unrecognized and in fact completely ignored till the turn of the century. In less than seventy years geneticists have produced a science of heredity that is one of the most solid, unified, and self-consistent bodies of knowledge that science has to offer. It is the biological counterpart of the great generalizations of physics: the interpretations of all motion by Newton's laws of mechanics and of all matter by the theory of atomic structure.

This undertaking was no mean task. In almost no field of nature is there such bewildering variety as in the types of reproductive devices exhibited by living organisms. Superficially, nothing may seem to differ more than do the reproductive mechanisms of very disparate organisms. The common molds can grow and reproduce indefinitely without sex and undergo sexual fusion only when two sexually appropriate types happen to meet. Most flowering plants depend on wind or on insect visitors to bring together sperm and egg cells. In fishes such as the trout and amphibians such as the frog the males fertilize the eggs outside the female body. Birds and mammals accomplish fertilizaton by sexual copulation. Even within a group of organisms, the range of different reproductive mechanisms can be surprisingly wide, as though in this realm natural selection had exercised an almost perverted imagination. Yet, beneath the bewildering variety of reproductive devices, genetics has unearthed the essential unifying regularities that occur in all sexually reproducing organisms. In so doing, genetics has atomized heredity into its elementary combinatorial components, the genes, and has found that the genes of all organisms consist of a common substance, truly the stuff that life is made of. The awkward chemical name of this substance, deoxyribonucleic acid, abbreviated into the symbol DNA, has become the almost mystical trigram that symbolizes life.

The story of genetics, and especially of the gene concept, is one of the most remarkable chapters in the history of the scientific enterprise. From Gregor Mendel and his garden peas to the American zoologist Thomas Hunt Morgan (1866–1945) and his fruit flies to the contemporary biologists James D. Watson and Francis Crick and the double helix of DNA, geneticists have unraveled the surprisingly simple rules that underlie the superficially bewildering complexities of heredity. They have interpreted those rules in terms of specific material objects, the genes. They have converted genetics from a purely biological science to a chemical one—molecular genetics—and have identified the individual genes with specific portions of nucleic acid molecules. They have deciphered the chemical script in which the instructions for the function of the genes are embodied, the devices by which the genes are copied when new cells are made, and the decoding apparatus that translates the chemical script of the genes into the chemical structure of proteins, which are the key products of the genes. Today man looks upon the specific materials of heredity, including his own, from the vantage point of a comprehensive, intellectually satisfying framework of knowledge. Future research will undoubtedly add new findings, but the basic structure of biology, resting on the twin foundations of evolution theory and molecular genetics, is here to stay, just as the basic framework of physics, resting on atomic theory, quantum mechanics, and relativity, is unshakably firm.

Molecular genetics—the analysis of the nature, function, and evolution of genes in chemical terms—was made possible by the earlier flowering of classical genetics—that is, the science of the transmission of inherited characters from one generation to the next. In turn, what made the successes of classical genetics possible was the discovery of a methodology that yielded clear-cut quantitative results suitable for mathematical analysis.

To achieve this, geneticists had to give up the attempt to infer the rules of genetics from the study of natural pedigrees of organisms such as domestic animals or man himself and focus their attention instead on experimental matings carried out purposefully on easily handled animals or plants, such as fruit fles or mice or maize, in the laboratory and in the experimental field. They had to work out the basic laws of heredity by studying simple cases—that is, crosses in which the two parents differed from each other in only one or a few well-defined traits, such as the color of a specific organ or its shape, or the presence or absence of some specific chemical reaction. They had to examine enough instances to yield statistically significant numerical results rather than merely qualitative observations. Only then was it possible to turn to more complex situations, such as the interpretation of human pedigrees in families with congenital diseases or the study of the inheritance of complex qualities such as body height or shape. In all cases, geneticists were able to show that the rules which apply to the simple situations apply also to the complex ones, with the difference that characters like body size or skin color which in man can range continuously over a wide scale of values reflect the inheritance not of one gene but of many. Body height in men and women, for example, is governed by a large number of genes that act at different times during the development of the individual, controlling the growth of his bones. A tall person has many genes that make for tallness and is likely to contribute several of them to his or her offspring. However, the actual height of an individual depends not only on his genes but also on the nourishment he receives during his growing years. In this as in other instances the genes set the intrinsic tendency of the organisms; the actual outcome is determined both by the genes and by the environment.

A single example illustrates the prototype kind of data on which the gene concept stands. A geneticist places together a group of males from an inbred colony of white rabbits and a group of brown females, also from an inbred colony. (The sexes could be reversed without changing the results.) The rabbits mate and have young. Those of the first generation are all brown. When these babies grow up and mate among themselves, one-fourth of their young are white and three-fourths are brown. The white character, which seemed to have disappeared in the first generation, has come forth again.

The explanation for the three-to-one ratio of brown to white is that each animal has two copies of the gene in question and passes one or the other of them to his or her offspring. The inbred brown rabbits have two copies of a "brown" gene, the inbred white rabbits have two of a "white" gene. All animals of the first generation, therefore, receive one brown and one white gene. They are brown because even a single copy of the brown gene causes the production of enough brown pigment to make the fur brown. The brown gene is *dominant* over the white one. In the second generation, only those rabbits that happened to receive a white gene from each parent are white.

Thousands of hereditary traits in hundreds of different species of plants and animals have now been studied, always with results of the same kind: the hereditary potentials that are called the genes are transmitted unchanged from generation to generation and behave like material elements, which are reproduced faithfully every time germ cells are made. There may be a hundred thousand different genes in man, somewhat fewer in smaller organisms. Every individual has two, and only two, copies of each gene, one copy received from the father, the other from the mother. Each cell of the body has, therefore, two complete sets of genes. When a cell prepares to divide, each gene reproduces, so that just before cell division there are four copies of each gene per cell. The mechanism of cell division, called *mitosis,* is arranged in such a way that, for each gene, each daughter cell receives only one copy of the paternal and only one copy of the maternal one.

When a new organism is to be created by sexual fertilization, something novel happens: the sex organs produce the germ cells, sperms and eggs. These have only one set of genes instead of two. Here the lottery takes place that decides, in a random fashion, which copy of a gene a sperm or egg gets—for example, in the experiment with rabbits the brown-coat or the white-coat gene. Then egg and sperm meet and fuse, and a new individual arises, again with two sets of genes in each cell.

Yet an organism does not necessarily manifest all the traits whose genes are present in its cells. Thus the brown-coat gene of rabbits, being dominant, masks the white-coat gene. Such dominance of one of the copies of a gene over the other provides one of the key mechanisms for evolution. When a new form of a gene arises by mutation, it is generally less useful to the organism than the normal, well-tried version of that same gene. If the new gene form were immediately expressed, it would probably handicap the individual organism and reduce reproductive success, so that the new gene would rapidly be eliminated. If, however, a mutated gene can remain hidden, so to speak, underneath a dominant gene for several generations while continuing to enter the germ cells, it has many chances of being tried in a number of new combinations with other gene sets. In some such combinations it may prove as valuable as or even more valuable than the original version of the gene, just as a human being may work effectively in a given environment and not in another. Thus new genes have a chance to establish themselves within a population and may ultimately even replace the original genes.

Not all mutated genes have such a chance. Some of them lead to the death of the organism as soon as they cease to be hidden under a dominant gene, either because they fail to carry out an indispensable function or because they produce some distortion incompatible with the over-all plan of development. Many of the deformities and monstrosities incompatible with life are due to such lethal genes, which can remain undetected until they happen to be present in two copies, without a dominant normal gene.

The essential feature of the genetic mechanism to emerge at this point is the fact that the inheritance of physical characteristics occurs by the orderly transmission from parents to offspring of material elements which remain unchanged from generation to generation. There is no blending or mixing of hereditary elements. There is no such thing as an inheritance by dilution as implied in such popular expressions as mixed-blood or pure-blood types. There is only the lottery of the innumerable combinations of genes in the egg and the sperm that generate a new individual.

If thousands of genes have to be distributed in a precise way every time a cell divides and every time eggs or sperms are made, a mechanism must exist to make this possible. This mechanism derives from the fact that genes are not loose in the cells but are fixed in the cell nucleus within microscopic filaments called *chromosomes*. A cell has two copies of each gene because it has two copies of each chromosome. Each cell of a human being has 23 pairs of chromosomes for a total of 46. One chromosome of each pair comes from the paternal sperm, the other from the maternal egg. A special pair of chromosomes called X and Y contains the genes that determine the sex of the individual. Women have two X chromosomes: hence all eggs have one X. Men have one X and one Y chromosome. Half of the sperms receive an X chromosome and when they join an egg give rise to XX pairs and therefore to females. The other half of the sperms have a Y chromosome and give rise to XY children, which are males. Occasionally something goes wrong and a man is born with one X chromosome and two Y's. Such XYY males do develop, but they tend to be abnormally tall and often are mentally retarded. There have been unconfirmed reports that XYY men may be prone to committing crimes of violence. Females with only one X and males with XXY are highly abnormal in bodily development and functions and are completely sterile. Evidently the X and Y chromosomes have genes whose precise quantitative balance is necessary for the organism to function well, especially in sexual development. Analogous situations hold for the other chromosomes. In fact, in man the absence of even a fragment of chromosome is generally lethal: if a germ cell happens to lose a chromosome (or a small piece of it), the embryo that results dies either in the womb or soon after birth.

Each chromosome contains thousands and thousands of genes, including many that are essential for life and normal development. At each generation the chance event that determines which one of each pair of chromosomes enters a given egg or sperm produces some reshuffling of the genetic material. If this was the only reshuffling, those genes that are in a given chromosome would remain together generation after generation. But the changes are much greater. As the sex organs prepare to make eggs and sperms, the two chromosomes of each pair line up together and exchange pieces, so that genes that were previously in the same chromosome now become separated. A given gene—for example, that for coat color in the rabbit—remains in the same relative position in the chromosome; what is reshuffled is the association of genes in a given chromosome. A rabbit may have the brown-coat gene in the same chromosome with, say, a short-tail gene. After an exchange, the short-tail gene may be in the same chromosome with the white-coat gene, and the new combination passes to the next generation.

The exchanges of parts between two chromosomes of a pair can take place anywhere along the length of the chromosomes. When an exchange takes place there is no gain or loss of genetic material, but only an exchange of equivalent parts of chromosomes. Such exchanges occur not only between genes but also within the genes themselves. If two defective versions of the same gene are present in the two chromosomes of an individual, exchanges within the genes may regenerate a normal gene in some of the germ cells. This is because the two defects were due to changes in different subunits or components of the gene, so that an exchange occurring between the two "bad" spots can reconstitute a "good" normal gene.

In this picture the chromosomes emerge as linear unbranched objects, in which the genes and their component parts are lined up in a way that is constant and characteristic. Every individual has each of the genes of its species in two copies. These can be different from each other, because of mutations that have occurred in the history of the species, as in the case of the white-coat and brown-coat genes of rabbits.

Different individuals of a given species may be unlike in many genetic respects. Two sisters or two brothers may differ in as many as 10 percent of their genes. The amount of genetic variation between different but related species is much greater. Direct tests are not feasible, since crosses between individuals of different species either cannot be performed or produce no live or no fertile progeny. Geneticists, however, have studied the chemical differences in certain proteins that seem to have persisted all the way from microbes to man and thus have learned something about the extent of evolution that has taken place in the corresponding genes. The results suggest that genes that control essential functions remain present and functional over hundreds of millions of years, during which time mutations accumulate within the genes at a steady pace, changing only those parts of the gene that are less critically needed for an effective function.

It is now reasonably well established that all differences between species are of the same nature as those within a species. They are determined by gene differences and are subject to the laws of genetics. The characteristic distribution of body hair that distinguishes man from other apes, his unique cranial capacity, the shape of his hand and fingers, the shape and color of wings and feathers that characterize the innumerable species of birds—all these traits, common to every member of a given species, are determined by sets of genes working together to direct the development of the body structure typical of that species. There is no evidence for, nor any need to postulate, the existence of special mechanisms different from genetic mutations and recombinations, by which new species might arise from existing ones. New species come into being when a population of a given species, because of geographic or other reasons, becomes genetically so different from the rest of the species that its members are unable to breed with those of the original species.

The knowledge of genes and their reshufflings is also relevant to simple organisms, like amoebae or bacteria, which produce copies of themselves by simple fission. The genetics of amoebae is poorly known because there are as yet no good methods of studying it. Bacteria, however, have genes and gene-strings that behave, at least at the molecular level, like the genes and chromosomes of plants and animals. In fact, most of the present knowledge of the structure of genes and of their functions has come from work on bacteria. Even viruses have genes and gene-reshuffling mechanisms basically similar to those of human germ cells.

When the genetic picture is considered as a whole, it becomes evident that the gene concept is not a simple one. On the one hand, a gene is an element whose presence in the chromosome controls some feature of the organism in a way revealed by genetic crosses. Thus the gene seems to behave like a unit. On the other hand, the gene is part of the linear structure of the chromosome, and this structure allows two copies of a given gene from different chromosomes to come together and exchange parts. The recognition that genes themselves are made of many linearly arranged subunits which can be reshuffled by recombination was a momentous advance in the history of genetics. It marked the transition from formal genetics—the study of the hereditary transmission of characteristics and of the corresponding determinant factors—to molecular genetics. The genes, previously viewed as theoretical constructs devised to interpret the formal rules of heredity, proved in fact to be subdivisible elements—that is, elements with a material structure that could be analysed.

There is an interesting similarity between the history of genetics and that of physics and of chemistry. The atom became a subject of study by physicists when it was understood to be, not an indivisible whole, but a complex of nucleus and electrons. Scientific chemistry began when molecules, the smallest amounts of substances that take part in chemical reactions, came to be seen as specific combinations of atoms. Likewise, molecular genetics started when the gene was recognized as subdivisible.

The phenomena of genetics, however, have one crucial feature that differentiates them from those of physics and chemistry. The properties of cells and organisms are controlled by elements —the genes—that are present in only one or two copies. The events that physicists and chemists study always reflect the average behavior of many individual elements, whose course becomes predictable through the law of large numbers. But genetics deals with phenomena that require, not the statistical presence of a certain amount of gene substance, but the deterministic, exact maintenance of all the millions of different genes in every cell. This is not accomplished by providing many copies of each gene to guard against statistical loss but by a rigid system which enforces the entry of just one copy of each gene into each germ cell and the exact duplication of each gene whenever a cell divides. This mechanism requires a degree of order, of organized information, that is found nowhere else in nature, not even in the movement of the planets.

It is a key feature of life to achieve the maintenance of a high degree of order in the face of the physical tendency of all organized systems to undergo increasing thermodynamic disorder. This achievement depends on the ability of living organisms to extract energy from their surroundings through a variety of chemical activities controlled by the genes. The ability of genes themselves to persist and to function, giving life its surprising capacity to continue and flourish, depends on the chemical properties of the remarkable substance DNA, which is the material substrate of the genes.

# 3 ∾ The Gene

The concept of the gene is at the center of biology. It forms the link between evolution and physiology, which is the study of how the genes of an organism perform in a given setting. The genes set the boundaries of that performance by determining what the cells of that organism can do. An altered gene may cause a disease. In man, gene differences affect, directly or indirectly, longevity, talents, and fertility. Since reproductive success determines the frequency with which the genes of an individual will be passed on to the next generations, the genes influence their own evolutionary success.

To arrive at a definition of a gene, it is necessary to review the properties of genes already mentioned, which any description of the gene must explain.

The first property is reproduction: a gene must reproduce in identical form at each cell division and, even when the structure of a gene is changed by mutation, the changed form must still be capable of reproduction. The second property is recombination: each gene must be able to pair precisely point by point with another copy of the same gene in a way that permits material exchanges between and within genes. The third property is function: each gene, present in one copy or two in a cell, must be capable of influencing the functioning of the cell; hence gene function must include some means of amplification.

These three properties already suggest for the gene a certain kind of structure. For genes to be copied exactly point by point, and to pair equally exactly point by point, requires an open configuration, in which the relevant details of structure stand exposed in order to be matched or copied. Only a line or a surface can so expose itself; a three-dimensional solid cannot. Thus genes can be expected to have a structure that is either one dimensional—that is, linear—or two dimensional. These are the classes of structures that can act as *templates* or molds on which new copies can be modeled.

In addition to the other properties, the structure of genes must account for their remarkable stability. Even apart from reproduction, genes retain their properties intact throughout the life of a cell. For human nerve cells, for example, this means practically the whole life span, because nerve cells persist from early infancy without dividing and without copying their genes. The only explanation of the stability of the genes is that they are molecules, held together by the same kinds of chemical bonds that bind the atoms of molecules such as those of water, alcohol, or sugar. These bonds are so strong that at normal temperatures they are hardly ever broken. The genes have a molecular structure: a structure that is common to all genes and at the same time allows each gene to have its own individuality, since each one must be unique, different from all the other genes in the same cell.

Obviously genes must be very large molecules, with enough atoms to permit a great variety of arrangements. The only very large organic molecules are *polymers,* in which many units (or *monomers*) of simple molecular structure are joined together in a linear sequence. There are repetitive and nonrepetitive polymers: the cellulose of plants is a repetitive polymer consisting of chains of identical glucose monomers successively linked to one another always in the same way. It is evident that such monotonous molecules could not serve as genes: they do not contain enough variety —or, information, as biologists prefer to say—to generate the enormous variety of different genes.

This concept of information is critical to an understanding of genes. A gene has a molecular structure which is unique and must be copied identically when the gene is duplicated. Copying requires that the gene acts as a template—that is, provides the point-by-point information for the copying—in the same way that a mold provides the information for the casting of a sculpture. In casting, the role of the mold is different from the role of a copying apparatus. The apparatus contributes general, nonspecific know-how, as for pouring metal and holding the mold in place. The mold contributes the unique, specific information which will make that sculpture and not any other—information embodied in the innumerable details of the template surface.

Polymeric molecules may serve as templates both for their own replication and for casting other molecules according to their directions. Two classes of polymeric substances present in all cells, the proteins and the nucleic acids, can provide the variety that the gene must have. Proteins and nucleic acids contain much more information than a substance like cellulose, because their molecules have many different units, attached to a uniform linear backbone, not in a monotonous repetitive order, but in an enormous variety of sequences. These sequences embody the information of the genes.

The sequence of units in these polymers may be compared with the sequence of letters in a written language. The twenty-six letters of the English alphabet, for example, can generate by their permutations a myriad of different words. Proteins and nucleic acids are language molecules. The alphabet of proteins consists of twenty different letters or monomers, which are called *amino acids*. The molecules of a protein, for example insulin or hemoglobin, consist of one or more linear chains of amino acids, usually with 50 to 1000 units per chain. All chains have a similar backbone; the different amino acids are ordered in different sequences in different proteins. In the analogy of language, the molecule of a given protein can be thought of as analogous to a word or a phrase. All molecules of a given protein are identical, so that a preparation of pure hemoglobin is like a series of millions or billions of identical words.

The nucleic acids are also language molecules, but their alphabet has only four units, called *nucleotides*. Each nucleotide consists of a base, a sugar, and phosphate. There are two kinds of nucleic acids, DNA and RNA, which differ in several respects. In DNA the sugar is deoxyribose, in RNA it is ribose. Sugar and phosphate make up the backbone, and the four bases attached to the backbone provide the alphabet for the nucleic acid language; in DNA they are adenine (A), guanine (G), cytosine (C), and thymine (T); RNA has uracil (U) instead of thymine.

Although both proteins and nucleic acids are language polymers, the amount of information they contain is obviously different. A four-letter alphabet like that of nucleic acids provides fewer combinations, fewer words, than the twenty-letter alphabet of proteins. A twenty-letter alphabet can generate $20^4$, or 160,000 four-letter sequences; a four-letter alphabet can only generate $4^4$, or 256. But this presents no difficulty: if the alphabet is poorer, variety is gained by using longer words. From four available letters $4^5$ or 1024 five-letter words, 4096 six-letter words, 16,000 seven-letter words, 64,000 eight-letter words, and over 200,000 nine-letters words can be made.

These considerations imply that the information of the genes —that is, the uniqueness of gene structure expressed in reproduction and in function—is specified by the sequence of the chemical units in a polymeric molecule. This surmise is correct.

Originally, proteins were thought to be the chemical substance of the genes, because nucleic acids were mistakenly regarded as repetitive polymers, in which ATGC or AUGC followed each other again and again in the same order. This idea was proved to be wrong, however, and DNA then became the leading contender for the role of genetic material because it is always found in chromosomes, and also because its amount is constant in all cells of a given organism, as one would expect the genetic materials to be.

A critical discovery was made in 1943, when the bacteriologist Oswald T. Avery (1877–1956) found that the DNA taken from certain bacteria could enter other bacterial cells and "transform" them—confer upon them some properties of the bacteria from which the DNA had come. Later it was learned that the entering DNA actually replaces the corresponding DNA of the recipient bacterium. A gene can thus enter a cell and replace the resident gene. Now it is recognized that the genes of all organisms, not only those of bacteria, are made of DNA. The only exceptions are certain viruses whose genes are made of RNA, the other type of nucleic acid.

Once the identity of genetic material has been established, the next step in understanding the gene is to clarify the structure of DNA molecules, to find out how they actually look and what is the arrangement of their atoms and groups of atoms in space. Since all functions of a molecule depend on its chemical reactions with other molecules, the functions of genes, including their reduplication every time a cell doubles, must reflect the arrangement of their atoms and the ability of these atoms to make specific contact with those of other molecules.

Unraveling the structure of a molecule cannot be done by looking, even with an electron microscope that magnifies 100,000 times. At best, the electron microscope will show the shape of large molecules: fibers of DNA or globules of protein. But the distances between atoms are of the order of a few billionths of an inch, beyond the power of the best electron microscope.

For simple compounds the task of determining the chemical structure and arrangement of atoms is not too hard: a series of determinations of atomic composition, of molecular weight, and of the presence of specifically reacting groups will generally tell the story. For large molecules, however, chemical analysis does not yield the total picture. In polymers like proteins and nucleic acids what is important is not only the arrangement of atomic groups in the polymeric chains, but also the arrangement of these chains in space. After the chains are completed they fold and associate in specific ways so that different parts of a chain or of different chains come together and join to produce the final molecular shape. The chemical bonds that cause the folding are sufficiently weak that the chains can unfold locally to permit changes of over-all structure. These giant molecules, therefore, are not hard solids but rather somewhat plastic, flexible solids, ready to open and close as needed.

To unravel the full three-dimensional arrangement of the

atoms of a large molecule is a laborious and demanding task. The most powerful approach is x-ray diffraction analysis; the pattern of reflections produced by a beam of x-rays that has passed through a crystal of a substance can reveal the arrangement of atoms and of groups of atoms in the molecules of that substance. If a crystal is not available, a fiber of the material under study may serve. Even with x-ray technique, aided by the use of high-speed computers to analyze the data, to unravel the detailed structure of a single protein such as hemoglobin or insulin often takes years, and in the end having learned the three-dimensional structure of one or two or ten proteins does not prove particularly helpful in understanding protein structure in general. Even though all proteins are made up of chains of amino acids joined in the same way, once these chains begin to fold and bend and curl up to form the maximum possible number of chemical bonds, they take shapes of an almost infinite variety. These shapes cannot be ignored because they provide the chemical surfaces that determine the function of each protein, either as a chemical catalyst or as a structural component of a cell. Ultimately, with the aid of high-speed computers, it will be possible to predict from the amino acid sequence of a protein its ultimate shape and chemical activities. But for the time being the protein chemist must still analyze proteins one by one.

Fortunately, the situation is simpler with DNA, the material of the genes. Almost all DNA has a common molecular structure, the famous double helix proposed by Watson and Crick: a structure amazingly simple and yet remarkably well suited to the roles of a genetic material. The understanding of this structure and of its implications for the reproduction and the function of the genes has been the key to the stupendous progress of molecular biology in the last twenty years.

DNA, as it exists in the chromosomes of cells, consists of two

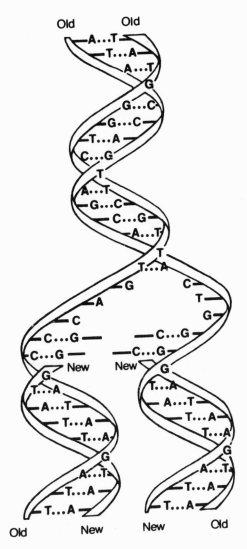

The double-helix structure of DNA and its replication

The ribbons represent the backbones of the DNA strands, constituted by alternating sugar and phosphate groups (not shown). The two strands are kept together by weak bonds ($\cdot\cdot\cdot$) between the nucleic acid bases A, G, T, C according to the pairing rules A$\cdot\cdot\cdot$T, G$\cdot\cdot\cdot$C. The lower part of the diagram shows what happens when the DNA double helix is replicated to generate two helices, the process moving upward. Note that each of the resulting double helices contains one old strand and one newly made one. (Adapted from James D. Watson, *Molecular Biology of the Gene*) polymeric fibers wound around each other in a helical way—the double helix (see diagram). As already mentioned, each fiber consists of a backbone of sugar-phosphate groups to which are attached the four bases A, T, G, and C in many different sequences. The two fibers that make the double helix are not independent of each other. The sequences of their bases are complementary, related by a specific matching of the bases A, G, T, and C. The rule is that every A in one fiber is matched and chemically bound to a T in the other fiber, and every G is matched to a C. The double helix, therefore, contains the information of its language symbols not once but twice, once on each fiber. If the sequence on one fiber is AATACGAG . . ., for example, the sequence on the other fiber must be TTATGCTC . . . The specific mutual attractions between A and T and between G and C hold the DNA double helix together, and they hold it together in precise molecular dimensions, because the chemical bonds between the pairs of bases A-T and G-C, which are called *hydrogen bonds,* have precise chemical lengths.

Hydrogen bonds, however, are weak bonds, much weaker than those that bind the atoms in the structure of the DNA fibers themselves. This fact has fundamental biological implications. The two fibers of a double helix, each of which carries the full amount of information of the genetic material, can separate relatively easily without causing the fibers themselves to break, and the separate fibers can either rejoin to each other or pair with any other fiber that has the appropriate complementary base sequence.

These properties of DNA immediately suggest the device for copying a DNA double helix: each fiber, as it separates from the other, makes available its sequence of bases, AACTGG . . . as a template on which a complementary fiber TTGACC . . . can be built by lining up the appropriate precursor monomers and then joining them into a fiber. As the diagram already presented shows, one double helix becomes two double helices, each indistinguishable from the original one. If the matching mechanism makes a chemical mistake—a mismatch—a mutation has occurred.

Notice that the role of the DNA fibers is to serve as templates only. They do not do the actual joining of the precursors to make new fibers. The function of a joining machine is performed by a set of catalysts, which work irrespective of the base sequences. As in sculpture casting, the roles of template and machine are separate, one providing the information, the other the know-how.

The principle of complementary pairing of nucleic-acid bases does not apply only to the copying of DNA. Each gene functions by directing the synthesis of what is called a messenger molecule —a fiber of RNA, which is complementary to one or the other of the two DNA fibers of that gene, but not to both. The DNA double helix opens up at the level of a specific gene or group of genes to make its base sequence available as a template on which the monomers for RNA can line up. A special set of catalysts joins these together to make the messenger RNA. Each RNA molecule, therefore, carries the information of those genes that served as templates for the synthesis. By analogy with the process by which a sequence of words is transferred from one kind of script to another—for example, from cursive writing to print—this process is called transcription, and the RNA messenger the transcript of the corresponding DNA. The messenger RNA then proceeds to the appropriate parts of the cell where it presides over the manufacturing of proteins. This principle of complementary chemical pairing also applies to the mechanism, to be discussed later, by which RNA guides the synthesis of proteins.

As far as is known, the principle of complementary pairing between sequences of nucleic acid bases is the only mechanism available for the specific transfer of information in living cells. This in essence is the so-called dogma of molecular biology: information flows from nucleic acid to protein, and not vice versa. The proteins are the final product of the genes. They serve as catalysts to perform the chemical business of the cell, and also in many other capacities, but cannot be used as templates to make more protein or more nucleic acid. In the making of proteins the information embodied in the sequence of bases of a nucleic acid is translated into the sequence of amino acids of a specific protein. But the information represented in the sequence of amino acids is not available for translation: it only serves to generate the shape of the protein and therefore its function. Informationally, proteins are dead-end molecules. The reason for this is that a template must be able to present its information in a readily usable form. The double helix of DNA, by opening up, lends itself to transfer of the information embodied in its linear sequence of symbols. RNA can do the same because it is a linear transcript of one DNA fiber. But proteins are different. Even though they have linear chains of amino acids in many assorted orders, these are assembled into complicated structures, unique for each protein. To make available the informational sequence of its amino acids a protein would have to unfold to a linear structure, a process that is reversible only very slowly and under special conditions. There are also other biochemical differences that make the use of proteins as templates most unlikely. It is not impossible that in early phases of evolution of life on earth some primitive proteins may have served to direct the evolution of nucleic acid genes by imparting to them their information, but if any such mechanisms existed they seem to have been lost.

Returning to the gene and the DNA, it is necessary to consider whether a gene is a DNA molecule. If by the word molecule is meant the chemist's definition of a separate particle that represents the smallest amount of a substance participating in a chemical reaction, the answer is no. The DNA extracted from cells consists of fibers of enormous length. A bacterium, for example, has 3000 or 4000 genes, all of which are part of a single double helix of DNA. Most viruses also have all their genes in a single nucleic acid molecule. In the cells of plants and animals each chromosome contains many DNA helices, each with hundreds or thousands of genes. This means that a given gene is a segment of a long stretch of DNA, just as a word is a segment of a sentence and a sentence is a segment of a book. From this another conclusion follows: within a DNA double helix the beginning and the end of each gene must be marked by special signals. But in DNA there are no signals except the sequence of the four bases; hence the special signals must be inscribed in this sequence. A gene, therefore, is a portion of a DNA filament that can generate a specific message and is, in addition, marked off at its ends by specific sequences of the A, T, G, C symbols. These signal sequences must direct the mechanisms that transcribe and translate the genetic message to start and stop at specified sites.

The signals that specify gene start and gene end play no role in DNA reproduction. The replication of the double helix to form two double helices starts at one point of the DNA fiber and proceeds, generally in both directions, until the ends are reached. In a living cell the duplication of DNA is always followed by cell division, which is so arranged that each of the two daughter cells receives one of each pair of DNA helices. In bacteria this simply means that one copy of the helix goes to each daughter cell. In cells containing many chromosomes, each with many helices, the complicated mechanism of cellular division called mitosis, not yet understood in molecular terms, insures that each daughter cell receives one copy of each helix. In mitosis, one copy of each chromosome is pulled by protein filaments to one of the two poles of the cell, and then the cell splits. The orderly distribution of DNA at cell division is obviously essential to maintain the constancy of the genetic endowment of cells. Experimentally, the distribution of DNA fibers can be followed by the use of substances containing radioactive atoms, which can be fed to cells in culture and become part of their DNA. When the cells divide, these atoms remain permanently in the DNA and by their radioactivity reveal that the individual DNA fibers remain unbroken and are distributed among daughter cells exactly as would be expected from the mode of duplication of DNA double helices.

Evidently the preservation of intact DNA is essential to life. In every cell, from bacteria to human cells, there are biochemical mechanisms that can repair the damages that rarely but inevitably befall the molecules of DNA, whether by errors of synthesis or by exposure to radiation or other hazards. These repair systems can rejoin broken DNA fibers, free adjacent bases that have been stuck together by the effects of radiation, or excise damaged pieces of one fiber and replace them by copying the opposite fiber. The occasional failure of normal repair may be an important source of mutations and therefore a significant factor in evolution. If the repair system does not function, the consequences can be serious. There is a congenital disease of man, called xeroderma pigmentosum, which is caused by a genetic defect in a DNA-repairing system. The affected individuals cannot tolerate exposure to strong sunlight; their skin cells are easily damaged and cancers of the skin develop frequently. They lack the normal mechanism, which would have repaired all or almost all of the DNA molecules damaged by the ultraviolet rays of sunlight.

Some viruses can be considered genetic oddballs because

their genes are made not of DNA but of RNA. A virus is an organism that multiplies inside the cells of other organisms, often producing disease. In their free form, the so-called virus particles, most viruses consist of one molecule of nucleic acid—DNA or RNA—wrapped in one or more coats of proteins. Some of the RNA-containing viruses—the virus of poliomyelitis among them—have a rather simple reproduction mechanism: the virus enters a suitable cell, makes an enzyme that copies the RNA of the virus, and makes the proteins needed to wrap the new viral RNA to make virus particles. As the cell dies, these particles emerge.

Other RNA viruses, however, go through a more complicated routine. The RNA enters the cell and makes a DNA strand complementary to the viral RNA. Here the complementary sequence device is used to pass the information from RNA to DNA, the opposite of the usual situation. Then the viral DNA can make DNA-DNA double helices, produce more viral RNA, and even install itself among the cell genes along with the chromosomes. The viruses that exhibit this transfer of information from RNA to DNA are responsible for producing leukemia and other cancers in animals. If the DNA made by the virus can find its way into the DNA of the cell chromosomes, it may from there determine the conversion of the cell from normal to cancerous one. Some scientists believe that certain spontaneous human cancers, especially leukemias, may be due to the activity of viral genes that after having remained buried in the chromosomes of human beings for many generations return to function under the impact of stimuli still unknown.

# 4 ∞ Genes in Action

A fertilized human egg gives rise to a complete adult organism, with millions or billions of cells of many different structures and functions. In some way the know-how of the genes guides the complex but precise process that generates the diversity of cells and organs and directs their functions and differentiation. An obvious explanation of this would be the assumption that different cells differ in the amount and quality of genetic material, that during differentiation various cells lose some specific portions of the set of genes that was present in the fertilized egg. However, this possibility has been disproved by experiment. All cells of an organism such as man have the same genes.

If so, then the differences between different cells must be differences in the functioning of the genes. Differentiation—the unfolding of the program of the organism inscribed in its genetic material—must consist of an orderly sequence of events regulating the activity of different genes. But before considering how this program is put into action, it is necessary to understand how genes work and how they impart to cells the instructions embodied in the sequence of chemical symbols of DNA. Interestingly enough, most of the existing knowledge of this subject has come, not from the study of complex organisms with many different types of cells, but from bacteria: not from studying how the same gene functions in different cells of one organism, but by observing how the genes of a bacterium respond to the needs of the organism when bacteria are placed in different surroundings.

DNA, as already explained, consists of two strands of nucleotides, whose four symbols A, G, T, C, following one another in all possible sequences, represent the information of the genes. Proteins have their linear information coded in the symbols of twenty different amino acids. Genetic and chemical tests have shown beyond any doubt that the linear map of a gene can be superimposed on the chemical map of a protein—that is, mutations that alter a series of linearly arranged sites on the map of a gene derived by genetic crosses bring about changes of amino acids located in a comparable order in the linear chain of the corresponding protein. Each gene, therefore, generates a protein chain by a transfer of information from the sequence of nucleotides to the sequence of amino acids. A mutation at one site in a gene usually changes one amino acid. It is now necessary to consider the remarkable set of devices by which, very early in evolution, this process of transferring information was perfected.

If each of the twenty letters of the protein alphabet is to be represented by symbols in the DNA alphabet, several symbols are needed. In DNA the symbols are the nucleotide bases A, G, T, C. Taken two by two, these nucleotides provide only $4^2$ or 16 permutations—not enough. At least three nucleotides are needed to code for each of twenty amino acids. Three nucleotides provide

51 &

$4^3$ or 64 permutations. Of the sixty-four possible triplets of nucleotides, sixty-one are actually used to represent instructions for amino acids. Thus most amino acids can be represented in the genes by more than one group of symbols, in the same way that the English word "man" can be translated into Latin as *vir* or *homo.* The set of nucleotide symbols corresponding to the various amino acids is called the *genetic code.*

Fifteen years of work have made it abundantly clear that, in all organisms investigated, the genetic code is a triplet code, in which three nucleotides, a *codon,* stand for one amino acid. The surprising fact is that the code, shown in the diagram, is the same in all organisms, from viruses to bacteria to man. It would be reasonable to expect that in billions of years the features of the code would have changed many times, that some of the features of the dictionary that translates the language of the gene into the language of the proteins would have evolved and been perfected, but this is not the case.

An interesting if partial explanation can be found in the fact that evolution tests organisms by the competence of their proteins as functional tools—mainly as catalysts or *enzymes* for chemical reactions. The functioning of each protein can be preserved, or destroyed, or occasionally improved by changes in one or a few amino acids. For the genetic code to evolve, a device should be found that would cause a DNA word to be translated into an amino acid different from the original one. But if such a device arose by mutation, not one but all proteins would be changed, not just at one point in their structure but at many. A change in the translating machinery would not have the same effects as a change in any one gene but would be equivalent to changes in every single gene. It would play havoc with all proteins of the organism, just as a defect in a printing machine would make an English-Latin

## The Genetic Code

Of sixty-four possible codons—that is, combinations of three symbols (nucleotides in the messenger RNA)—sixty-one correspond to amino acids. Some amino acids have only one codon, others as many as six. For each codon there is one amino acid adapter with an appropriate anticodon. The signal for initiation of protein synthesis is a partially known sequence of symbols ending with AUG, which is the symbol for methionine; hence all proteins when first made start with methionine. The three remaining codons are termination signals, any one of which causes the protein chain to come off the synthesizing apparatus.

| Amino Acid | Codons on RNA |
| --- | --- |
| phenylalanine | UUU, UUC |
| serine | UCU, UCC, UCA, UCG, AGU, AGC |
| leucine | UUA, UUG, CUU, CUC, CUA, CUG |
| tyrosine | UAU, UAC |
| cysteine | UGU, UGC |
| tryptophan | UGG |
| proline | CCU, CCC, CCA, CCG |
| histidine | CAU, CAC |
| glutamine | CAA, CAG |
| arginine | CGU, CGC, CGA, CGG, AGA, AGG |
| isoleucine | AUU, AUC, AUG |
| threonine | ACU, ACC, ACA, ACG |
| asparagine | AAU, AAC |
| lysine | AAA, AAG |
| methionine | AUG |
| valine | GUU, GUC, GUA, GUG |
| alanine | GCU, GCC, GCA, GCG |
| aspartic acid | GAU, GAC |
| glutamic acid | GAA, GAG |
| glycine | GGU, GGC, GGA, GGG |
| INITIATION | [..?.. AUG] |
| TERMINATION | UAA, UAG, UGA |

dictionary utterly useless. No enzyme would work, and the whole structure of the organism would disintegrate. Mutations that changed the translation mechanism would be lethal. The catastrophic consequences of such mutations must in fact have been the key to the conservatism of the genetic code, from the early organisms all the way to the living things of today.

Sixty-one of the possible codons in DNA stand for amino acids. But not only are codons for amino acids necessary; start and stop signals are needed as well, to make protein chains that begin and end at precise points located in the middle of the nucleic acid template. The stop symbols—the periods in the protein language—are the three remaining triplets at the sixty-four—the ones that correspond to none of the amino acids. Like the rest of the code, these stop symbols are universally the same. The start signal is likewise universal, but its detailed sequence is more complicated and not yet fully understood.

The story of how geneticists and biochemists discovered the over-all features and the actual details of the genetic code is as fascinating as the reconstruction of an ancient language from a few surviving documents and tablets; more fascinating, if anything, because the unraveling of the genetic code required the convergence of many different approaches that would hardly seem related to one another—from the purification and chemical analysis of viruses, to the isolation of enzymes that synthesized abnormal nucleic acid, to years of genetic work on one single gene of a *bacteriophage* (a virus that attacks bacteria), to decades of studies on the synthesis of protein in the test tube. Even more important than the genetic code itself, this work has provided a fairly complete and detailed picture of the workings of the genes and, incidentally, a foundation for future approaches to the direct correction of genetic diseases.

As mentioned earlier, the DNA of the genes is not itself the template that lines up the amino acids for the synthesis of proteins. The intermediary template, the messenger of the gene, is a strand of RNA transcribed off one and only one of the two strands of DNA.

All transcription of DNA to make messenger RNA (and also some other RNA molecules that serve other functions) is done by an enzyme called RNA polymerase. The entire function of the genes depends on this enzyme, a central link in the business of life. Transferring responsibility for template action from DNA to RNA makes it possible to have proteins made in any part of a cell, not only next to the chromosomes. More important, it provides amplification, since RNA polymerase can make many transcripts of a gene without gene reproduction. This makes it possible to regulate the function of individual genes by regulating the number of transcripts of a given gene made by the RNA polymerase in different cells or in different environments.

Once the RNA polymerase starts transcribing a segment of DNA, it proceeds at a constant velocity, which is relatively slow, in fact adding about thirty nucleotides per second to the growing RNA molecule. It stops only when it encounters some signal—still unknown—that indicates the termination point for that messenger. This signal is not the same as those that indicate the end of a gene: a single messenger molecule can be longer than one gene and may in fact serve as messenger for one, or two, or as many as ten or twenty genes. Since, once started, the making of messenger molecules continues at a constant speed, any regulatory mechanisms must act at the critical moment when the RNA polymerase starts making the RNA transcript of a gene or a group of genes.

In the next step, in which messenger RNA directs the assem-

**The synthesis of RNA**

The enzyme RNA polymerase attaches itself to certain locations on the DNA fiber, causes the two DNA strands to separate, and proceeds to synthesize an RNA molecule using one DNA strand as the directing template. Precursor molecules (not shown) contribute the nucleotides which the enzyme adds one by one to the growing RNA fiber. After the enzyme moves on, the DNA strands come together again and the RNA strand becomes free. The pairing rules for the DNA-RNA strands are A with U, G with C. Compare the pairing rules for DNA strands. (Adapted from James D. Watson, *Molecular Biology of the Gene*)

bly of the corresponding protein or proteins, a novel problem is encountered—the means by which a sequence of three nucleotides in the messenger physically recognizes the corresponding amino acid. It does not combine with it directly: all specific pairing is always between nucleotides. What happens is that each amino acid, prior to its incorporation into a protein, becomes attached to an "adapter" by the action of a specific enzyme that works on that amino acid and that one only. The adapter is a small molecule of RNA; there are as many adapters as there are codons in the genetic code. An adapter molecule is shaped more or less like a hairpin. At the rounded end it has a group of three nucleotides, the *anticodon,* matching the corresponding codon, as shown in the diagram. At the free end the adapter has the proper amino acid

RNA polymerase

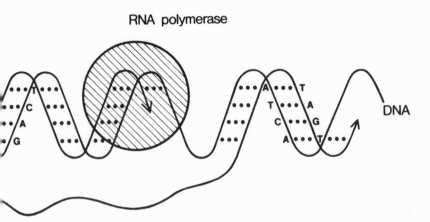

DNA

attached. In this whole process the only specific function beyond the pairing of nucleotides is carried out by the enzyme that attaches a given amino acid to its appropriate adapter. After that, all is smooth sailing: the anticodon on the adapter will pair with the appropriate codon of the messenger RNA, lining up the amino acid in its proper place.

This process does not take place automatically: it requires the intervention of an extremely complicated mechanism. But this mechanism does not make any contribution to the specificity of the process, the translation itself. It is like the typewriter, which is an essential but unoriginal part of any translating machine. It serves to line up messenger and adapter in the proper position, it helps recognize start and stop signals, slide the messenger along like tape in a tape recorder, and seal together the amino acids into protein chains.

Once a protein chain is made a number of things can happen to it. A common change is for one or more amino acids to be snipped off from the "start" side. Sometimes the reshaping is more

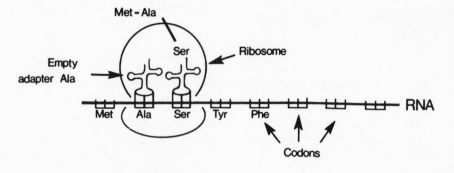

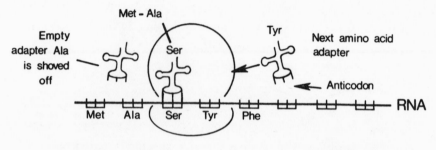

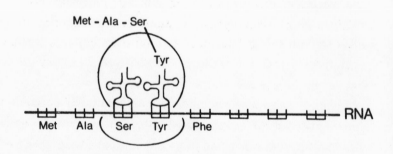

## The synthesis of protein by elongation of amino acid chains

Only one stage in the enormously complex series of events is represented here. The RNA template with its triplet codons becomes associated with a structure called a ribosome, which serves to line up all the components of the system. Synthesis occurs from left to right in the diagram. On top, the growing protein chain Met-Ala has just gained the next amino acid, Ser, which is still held on its clover-shaped "adapter." In the center section, the empty adapter falls off, the ribosome shifts one codon down, and the next adapter with the amino acid Tyr comes into place. At bottom, the protein chain gains the amino acid Tyr, whose adapter now carries the entire chain. The process continues till a termination codon on the RNA causes the protein chain to fall off the last adapter. The adapter molecules are made of RNA, each having an anticodon triplet (for example, GCA) to pair with the corresponding codon (in this case, CGU). (Adapted from Albert Lehninger, *Biochemistry*) drastic. A piece of the chain may be cut from one end, or from the middle, as in the case of the hormone insulin. Sometimes a very long protein chain is made and then split into many pieces: this is what happens to the polio virus. How general such devices are in living cells is not known. Only a few proteins have been synthesized in the test tube, and much remains to be learned. In any case, once completed and sometimes reorganized by various cuts, each protein proceeds, by the folding of its amino acid chains, to take the specific shape on which its function depends.

It is now possible to explain how the specific functioning of the different genes is regulated.

Beginning with RNA polymerase, the enzyme that transcribes the RNA messages off the genes, it is evident that changes in this enzyme could, by altering its specificity, modulate the effectiveness of transcription of different genes or sets of genes. This does in fact happen—for example, when a bacterium being deprived of ample food begins to burn its reserves and changes itself into a spore, a dormant cell that weathers days or years or millennia of dryness and food shortage. The shift from making normal cell materials to making spores involves a change in the structure of RNA polymerase that alters its ability to transcribe different genes. A similar process of regulation by changing this enzyme also happens during the infection of some cells by viruses. An alternative possibility would be for an organism to have not one but several different RNA polymerases specific for different sets of genes and to adapt itself to changing needs by turning on or off one or another of these enzymes.

The best-known mode of regulation in bacteria is the regulation of the function of a given gene or group of adjacent genes by modulating the frequency with which the transcription process starts. Modulation of gene functions in living organism must be both powerful and precise. A nerve cell makes no measurable amounts of muscle proteins, and conversely muscle cells make no nerve proteins. If a bacterium is grown in meat broth, it makes almost none of the enzymes needed to split and use milk sugars. A few minutes after milk sugar is added to the broth, the sugar-splitting enzymes start being made at the full rate, about 1000 times the rate in plain broth. The mechanism is as follows: There is in the bacterial cell a gene that makes a *repressor* protein specific for the milk-sugar enzymes. This combines with the DNA of the corresponding genes near the point where RNA polymerase would start making the messenger for milk-sugar enzymes. As long as the repressor is on the DNA, the polymerase cannot start making that messenger. Milk sugar combines with the repressor and causes it to come off the DNA so that messenger and enzymes can be made. When the milk sugar is all used up, the repressor goes back onto the DNA and the synthesis of the milk-sugar enzymes ends.

Such a device does not produce only on and off signals. The concentration of milk sugar in the broth gears with great precision the level of enzyme made to the level of enzyme needed, by controlling the proportion of time during which molecules of repressor stay on or off the DNA. Thus an accurate feedback system is established, which prevents the bacterium from making any more enzyme than it needs. A mutant bacterium lacking the repressor for milk-sugar enzymes can survive, but since it continues to make these enzymes for which it has no use, devoting 3 percent of its total proteins to this futile endeavor, it grows 3 percent less fast —a difference that in nature causes it to disappear in a few hundred generations of competition with normal bacteria.

The milk-sugar case is only one example of a general situation: in bacteria, almost every biochemical process is similarly regulated. If a bacterium that was making all its amino acids starts receiving food that contains one of those amino acids, it immediately stops making the enzymes that catalyze the synthesis of precisely that amino acid. Here again a repressor protein exists, which, in the presence of an excess of the amino acid, combines with the DNA at the side of the appropriate genes and stops their transcription. Sometimes the mechanism is not a repressor: a regulatory protein may help the RNA polymerase attach to a specific gene, and this activating protein may be functional only in the presence of the signal substance. Some regulatory mechanisms reflect preference between different foods: a readily usable sugar may promote the production of a regulator substance that keeps the cell from making the enzymes that would handle other sugars. Even more rapid regulation is accomplished when an enzyme needed to make an amino acid is inhibited by increased levels of that amino acid—a typical case of feedback control.

Thirty years ago the humble bacteria were believed by some chemists to be merely bags of catalysts and their substrates, practicing a set of textbook exercises in elementary physical chemistry. These naive chemists were vastly mistaken. The bacterial cells have proved to be the true virtuosi of the regulatory device. The sophistication of their self-adjusting chemical network challenges the most expert computer programmer. Bacteria as cybernetic systems, poised for maximal efficiency, bear witness to the marvels that natural selection can accomplish by perfecting the controlled functions of the genes.

From the standpoint of biochemical regulation, the evolution of higher organisms may have been much more complicated and demanding. Premium may have been placed on ability to perform specialized tasks in a more or less constant environment like that of the human body, rather than on the prompt and efficient adaptation to rapid changes in chemical surroundings that were important to bacteria. This may explain why the regulatory mechanisms that are so strikingly effective in bacteria have not been found in animal cells in comparable form. The devices that regulate the functioning of genes in animal cells are still a mystery. One clue is given by the fact that messenger RNA molecules, which in bacteria are very unstable and are broken down after having been translated a few dozen times at most, are much more stable in animal cells. In young red blood cells, hemoglobin may continue to be made on the same messenger molecules for at least several days. In this case, shutting off the production of messenger would yield no great saving of protein. A more efficient regulation would be to control the translation of RNA information into protein. That is the level at which some hormones may work.

Another important regulatory mechanism is the control of DNA replication. When cells of plants or animals undergo cycles of division, they go through a precise and orderly sequence of events. For a long while after a mitotic division no DNA is synthesized; then within a few hours the amount of DNA doubles and each chromosome has two copies of each gene. Another short period without DNA synthesis follows, after which cell division takes place.

The master switch of this cycle is apparently the DNA replication itself. Whenever some stimulus triggers the synthesis of DNA, a complete cycle follows, including complete doubling of DNA and, a few hours later, cell division. As in any cyclical program, one stimulus is sufficient to set the whole cycle going. Most mature cells of the human body cease to divide unless some unusual accident happens. The trigger must somehow have become inhibited in the course of development. The reason cancer cells divide and damage the body may be that they have become insensitive to the unknown trigger-restraining mechanism.

As for the replication of the individual DNA molecules, most of the information again has been gained from bacteria, which have all their genes in single file on one circular DNA fiber. Replication starts at one point, always the same one, and proceeds in both directions, so that when the two points of replication meet, one circular fiber will have generated two fibers. Here too the cell-division mechanism is precisely geared to DNA synthesis and again appears to be triggered by an event that takes place when DNA replication starts. The nature of the tie between DNA synthesis and cell division, in bacteria or in other cells, remains unknown. Ultimately the signal that starts the copying of DNA must directly or indirectly influence the cell membrane, which is where the events of cellular division take place, but the process is not understood.

One thing, however, is certain. The astonishing precision of function by which the genes respond to the needs of the cell and of the organism is not superimposed onto them by some other agency. It is inherent in the genes themselves, their structure, the structure of the substances that they produce. Repressors and activators, as well as all the components of the machinery for copying and transcribing the genes and translating their information into the structure of proteins, are primary products of individual genes. This system is not static—it is continuously being tested and modified by the action of natural selection.

The harmony of the genes has something of the same grandeur as the harmony of the heavenly spheres, but with the difference that the harmony of the genes is not immutable. It is rather a flowing chorale, superbly adapted to the present, yet evolving to remain in tune with the uncertain future.

# 5 ഗ Cells

The way the hereditary messages of the genes are translated into the structure of the proteins, which are the working molecules of living cells, has thus far been considered as though these molecular events occurred in isolation. In fact, they can be studied in the test tube by mixing the appropriate template molecules, DNA or RNA, with the appropriate catalysts and the precursor substances that provide the building blocks for the *macromolecules* to be synthesized including the nucleic acids themselves as well as proteins. While the copying of the DNA of the genes is the essential process in securing the continuity of life's substance, the production of RNA and proteins is also indissolubly bound to that continuity because they provide the machinery for making more genes.

The next step is to learn how these processes take place; what kinds of substances the cell uses as precursors for the macromolecular substances and why; and how these substances are prepared from the food received by the organism. In other words, molecular biology must be related to processes of everyday life: nutrition, food-calories, decay. But first another basic aspect of biology must be considered.

A biochemist studies the reactions of life in a test tube in which he mixes whatever chemicals he finds necessary to make reactions take place. But in the organism the biochemical reactions occur in a very unusual environment: the living cell. All cells have certain basic features in common. Ranging in size from the smallest bacterium, of which a thousand billions would fit into a thimble, to the largest nerve cells, almost visible to the naked eye, with fibers several feet long, all cells provide environments superbly appropriate for the tasks to be performed by their genes and gene products.

Practically all organisms living today consist of cells. The only exceptions are the viruses, and they are the exception that confirms the rule. A free virus—for example, a particle of polio or influenza virus—is simply a piece of nucleic acid surrounded by one or more protective layers of inert substances. In that form the virus is inactive; it is dormant. Before it can function and multiply it must meet a cell and inject into it its nucleic acid—that is, its genes. Only when they are inside a cell can the genes of the virus function and be copied.

The cell provides the conditions that permit the effective performance of the chemistry of life. Essentially, it is a chemical factory. It has a power plant that provides the energy needed for chemical transformation. It has partitions that segregate various pieces of chemical machinery, keeping them from damaging one another. For example, catalysts that can break down proteins are kept apart from the functional cell proteins and are directed to the outside of the cell where external proteins are to be used for food. Each cell has an outer membrane that regulates the concentration of chemicals within the cell. This membrane is neither a solid wall nor a sieve; it is an active and selective device that can recognize individual substances and control their flow in and out of the cell. Its basic structure is that of a double layer of fatty molecules with protein molecules embedded among them. By the specific properties of these proteins, the membrane regulates the passage of various chemical compounds on the basis of the size, shape, and electrical charge of their molecules; it can actually pump a substance in or out of the cell depending on the needs. Because of such pumps, a substance may be a thousand times more concentrated inside a cell than outside, and its level may be maintained at a precise value, or alternatively changed in the direction needed for optimal function. Thus the transmission of impulses in the nerves depends on the functioning of a membrane pump that regulates the movement of electrically charged sodium and potassium atoms in and out of the nerve fibers, altering the ratio of internal and external electrical charges.

A cell, therefore, must be regarded as a circumscribed domain in which the processes of life operate in a chemical environment that is kept nearly optimal by a combination of several mechanisms. The internal synthesis of various compounds is precisely regulated by the functioning of genes and enzymes; the entry and exit of various substances are regulated by the properties of the membrane and its proteins. In this sensitively controlled chemical environment the building of cell materials and other chemical functions of the cell take place with an efficiency that is often greater than that of the most carefully designed machine.

It must have been the superb efficiency of the cellular organization that made it supersede any precellular forms of life: an inefficient system would not be expected to have persisted in competition with a superior one. Evolution, however, has not made cellular organization uniform. Even today the existing range of organisms exhibits two main types of cellular organization: one found in bacteria and in some seaweeds; the other in plants and animals. This suggests that in the course of evolution two basic plans of cellular organization have each proved to be so well suited to its own task in its own environment that they have spread and persisted, each occupying a substantial sector of the biological world.

The cellular organization of bacteria is called *procaryotic*—that is, antecedent to a nucleus—and is relatively simple. The electron microscope reveals an outer membrane but no internal compartments. There is no nuclear membrane, no chromosomes, no mitosis. All the genes of a bacterium are part of a single DNA double helix, which if stretched out would be a thousand times as long as the cell. When the cell divides, one newly made copy of the double helix goes to each daughter cell. The absence of inner membranes means that various substances within the cell are not compartmentalized. They are not mixed at random, however. Molecules of different substances, especially protein molecules, have many ways of associating themselves in specific complexes for optimal functioning, just as machine tools are arranged in a production line for optimal speed and efficiency of output. Certain proteins, as has been mentioned, are located specifically in the cell membrane to regulate the transport of specific substances. One of the fascinating problems of cell biology is how a protein finds its own place within the fatty layers of a membrane or, more generally, in an organized molecular assembly.

The other type of cell, which is called *eucaryotic*—well nucleated—and has been adopted all the way from yeasts and molds to the higher plants and animals, is much more elaborate. The genes are organized in chromosomes within a well-defined nuclear membrane, and each chromosome has many double helices of DNA with thousands of genes each. Outside the nucleus the cell body is subdivided by membranes into many compartments: some contain the catalysts that serve to digest food, others eliminate food residues, others generate chemicals needed for special purposes.

In view of the fact that each organism has only one or the other of the two types of cells but not both, it would seem reasonable to expect that the lines of descent of the two classes of organisms evolved independently. There are, however, enough similarities—from the structure of the DNA to the genetic code to the basic plan for protein synthesis and structure—to prove that the two types must have had an ancestry and early history in common. At some time the two diverged. Probably the eucaryotic type evolved from a simpler procaryotic ancestor and proved more suitable to the building of complex organisms with many different cells with specialized functions.

But surprises are always in store for the biologist: evolution seems to have tried all possible twists in its quest for the successful. All eucaryotic cells contain some membrane-enveloped vesicles, smaller than the cell nucleus and highly specialized in function: these are the *mitochondria,* which function as the power plants of the cell when oxygen is present. They take chemicals derived from food and oxidize them—that is, they remove some of their electrons and transfer them to oxygen to generate water. This chemical process releases a great deal of energy, which the mitochondria store in the form of a chemical called ATP. The ATP coming from the mitochondria is then used in all sorts of chemical reactions—synthesis of proteins, nucleic acid, fats and sugars—as well as for cell movement and muscular contraction. Cells that use a great deal of energy very fast, such as those in the flight muscles of birds, have enormous amounts of mitochondria.

What the newest studies have discovered is that the mitochondria of eucaryotic cells resemble in some fundamental respects the cells of procaryotic organisms. Each mitochondrion, about the size of a small bacterium, has a small piece of DNA and a machinery that can use the genes of that DNA to make a few proteins, and this machinery resembles that of bacteria rather than that of the rest of the eucaryotic cell. The mitochondria might tentatively be described as remnants of bacterial cells installed in eucaryotic cells in a symbiotic relation—that is, a relation of mutual benefit. What happened in evolution can only be conjectured. Cells of an evolving line, possibly amoebalike, unable to use oxygen, may have ingested bacteria. Occasionally some of these bacteria did survive and by using oxygen and producing ATP contributed to the reproductive success of the host amoeba. It would be an enormous advantage for the amoeba to extract ten or more times as much usable energy from food by the use of oxygen. Thus the symbiosis would have given rise to a new and very successful line of descent, committed to the persistence of the symbiotic association: in time, the bacteria would become mitochondria. Reproducing inside other cells, they would come to need less and less independent chemical machinery. Any mutation causing loss of an enzyme that was no longer needed would not be damaging but actually helpful, since it would prevent the mitochondrion from making a protein it had no more use for. When cloth is all bought in stores, there is no need for a loom in the home and the space may profitably be used for something else. The process of loss of unneeded functions is very common in evolution: worms parasitic in the intestine regress to simpler forms; fish in underground lakes lose their eyesight and often their eyes altogether. Mammals, including man, having become adapted to eating organic food, have lost the ability to manufacture many organic chemicals that plants and bacteria can readily make.

In addition to differences and possible relationships between the procaryotic and eucaryotic cells, the diversities between cells within a single complex organism are also instructive. Apart from the over-all plan of the eucaryotic cell, which is so uniform that nucleus, membranes, and mitochondria in all cells of different organs can be recognized under the microscope, the differences between cells are remarkably pronounced. Consider the disparities between the cells of muscle, filled with contractile material, those of the brain with their long fiber, and the red blood cells filled with hemoglobin. In fact, a cell of a given type in man resembles an analogous cell in a fish or a frog or a fly more than it resembles a cell from a different human organ. The reason for the specialization is that the function to be accomplished determines the structure of the cell that accomplishes it. This is an essential feature of differentiation: specialization by changes in the functional organization of cells with the same genes but with different functions.

The differences between cells reflect the presence of different proteins. Since proteins are products of genes and all cells of an organism have the same genes, different genes must function differently in different cells. Not that a given gene makes a different protein in different cells: each gene makes a certain amount of its own protein in each cell type. Thus in the red blood cell over 90 percent of the protein is hemoglobin, which is the product of only two genes. Muscle cells make no hemoglobin but make many other proteins, including enormous amounts of two that together form the contracting element. The differences between cells in different organs extend to reproductive activity. Nerve cells in the brain stop dividing very early; no more new cells are made in children past the second year. The existing ones persist and live till old age, with no DNA synthesis and no gene reproduction. The cells that line the skin or the intestines, however, continue to multiply and to die all the time, so that these organs have a perpetually renovated population of short-lived cells. The cells of the reproductive organs that generate eggs and sperms function at very different rates in different periods of life: in women all eggs are present at birth and mature one a month after puberty; in men the sperm precursor cells multiply and mature from puberty to old age.

Thus superimposed on the general plan of the eucaryotic cell there are structural differentiations that reflect the task assigned to each cell type in the plan of the organism. Presumably the organizational flexibility and elaborate structure of the eucaryotic cell permitted it to generate the immense variety of specialized cell types needed as the whole range of multicellular organisms evolved. The procaryotic cell apparently did not have enough "differentiation potential" to evolve into complex organisms, although what it lacked in differentiation potential it made up for in biochemical potential. The cells of animals are very exacting in their food requirements. Living in the sheltered environment of the body, fed by blood and other body fluids that maintain a fairly constant level of nutritive substances, these cells have in the course of evolution become dependent on an external supply of amino acids, vitamins, and other substances to build their proteins and nucleic acids. In contrast, plants have simple but rather specialized requirements: they can manufacture all the organic compounds they need. They use carbon dioxide, water, and light to make sugar and to extract nitrate, phosphate, and other salts from the soil.

Bacteria are different, not only from both plants and animals but even from one another and can be even more specialized than human cells. Some of them cannot make any of the simple substances needed to build their cell material and, in order to grow, must derive all these substances from the environment. Such bacteria have become specialized to a life in substances such as milk or wine, which provide them with a wealth of organic nutrients. Other bacteria can live and grow with nothing more than water, sugar, and a few salts. From these few substances they make what they need to make more cells, including proteins, RNA, and DNA. Besides being easy to please, these bacteria are also amazingly flexible in their diet. When they run out of their favorite food, say sugar, they can use any one of many other organic compounds, including the most improbable foods like camphor or gasoline or polystyrene. Other bacteria thrive on hydrogen sulfide plus carbon dioxide, and one group can use as food carbon monoxide, the poisonous gas in the exhaust fumes of automobiles. The shift from one kind of food to another is accompanied by amazingly precise rearrangements of the chemical machinery, almost computerlike in their mode of operation.

Cells, then, exhibit many differences in structure and function, even though they all have genes made of the same substance, DNA, which use the same genetic code to direct the formation of proteins by a universally common process. The differences are either intrinsic to the evolutionary type—procaryotic or eucaryotic—or reflect differences in the genetic endowment of different organisms, or result from different specializations of gene functions within the same organism. Even when cells are part of a physically rigid structure and contribute to the over-all form of an organism, their role is seldom passive, like that of building stones, but generally is active. The rigid structure of bone, for example, is a product of the chemical activity of bone cells, which are imbedded within the bone and build and rebuild it all the time as the body requires.

With bacteria, a single organism grows to a certain size, divides, and produces two bacteria, and the process continues until the food runs out. The key phenomenon—one is almost tempted to say the goal of the growth process—is the exact duplication of the set of genes that makes the two daughter cells genetically identical to the mother cell. In a complex organism such as man, the tasks have become specialized. Most cells grow and divide, always duplicating their gene sets; others stop growing and remain unchanged through the life of the organism. The continuity of the life of the species is delegated to the special cells that produce the generative cells. The rest of the organism is designed to protect and feed and bring to fruition the genetic material of eggs and sperm. From the point of view of evolution, the maintenance of the exact amount of gene material matters only with regard to the sex cells. Other cells of a complex organism might safely lose genes or chromosomes if this led to more effective function. In fact, there are organisms in which changes in chromosome numbers take place in the course of development, but these are rare. Either there must be some advantage in having all genes in all cells of the body, even if some genes are not needed, or else the mechanism of mitosis that partitions chromosomes exactly at each cell division was too valuable to be tampered with for the sake of altering the number of chromosomes during development.

I have sometimes referred to evolution—that is, natural selection—almost as if it worked with a goal in mind. However, it is important to remember that there is no goal, or, more specifically, no goal of duplicating genes or preserving the species. Evolution does not work on genes or on species but on organisms. The cell has evolved as organism: either as a single-cell organism whose descendants were more or less successful depending on the functioning of their genes in their environment, or as part of complex organisms whose millions and billions of cells assumed specialized functions, again dependent on their genes and tested in specific environments. In describing these events there is a tendency to use such words as goal or purpose or plan because one comes to think of genes or cells or organisms in human terms. But genes are merely molecular structures, cells chemical factories, and organisms the pawns of the blind master, evolution. Man is alone in knowing the joys and torments of conscious will.

# 6 ஒ Energy

In the morning before I go jogging I drink a glass of orange juice. It provides water and energy. Why I need water is obvious: I shall sweat at least that much water while running. What is less obvious is why I need energy and how I use it, and how the energy stored in the sugar of orange juice is made available to the cells of my body.

One way people use energy is in performing mechanical work with their muscles. Likewise, bacteria or other microbes that move around by shaking or contracting their surface appendages use energy to do so. In most cases, however, movement does not create the principal demand for energy. The greater part of the energy used by an organism is expended to carry out chemical syntheses. This is a key concept which should be made as clear as possible. When a cell uses a mixture of twenty different amino acids to synthesize all the proteins it needs, or a mixture of precursor molecules to make nucleic acids, or of sugars to make sugar polymers, and so on, it joins them together in chains, removing one molecule of water for every unit it adds to the chain. The joining process requires energy. If one gram of protein is burned, it releases more heat than the burning of an equivalent amount of the proper mixture of amino acids. This means that the polymer has stored more energy in its chemical bonds; this energy of biosynthesis must have come from somewhere.

Chemical reactions always proceed in the direction that reduces the total amount of available energy. Proteins, therefore, should break down into amino acids, not the reverse. This is in fact what happens. If amino acids are mixed together they do not combine to make proteins: if anything, proteins left in water have a tendency to break down to amino acids. This happens very slowly, however, both in the test tube and in living cells. The biochemist working with proteins reduces the breakdown by freezing protein solutions to store them. Since cells cannot do this, they lose a small percentage of their protein molecules each day.

It is understandable that to assemble large molecules from small ones should require energy. There is a general law of physics which states that changes in a system always increase the total disorder of the system; this is the *entropy law*. Increasing the temperature increases disorder because molecules move around faster. The synthesis of a polymer from its components represents a local increase in order and, therefore, must be compensated for by a greater increase in disorder somewhere else. The over-all increase in disorder means that some energy has been dissipated.

For molecules like proteins or nucleic acids, whose sequence of units is dictated by another molecule acting as a template, the template might be thought to supply some of the order gained in synthesis. But the template only influences the specific sequence of units, not the process of joining them together. Most of the increased order resides in the fact that the joined monomers are not as free to move around as they were when uncombined. Both bond energy and order are gained in polymerization.

Gains in energy come from the chemical energy of foodstuffs, including the sugar in my orange juice. Chemical energy is the only form of energy (except that of sunlight captured by plants) that organisms can make use of. In essence the process is simple: the energy stored in the chemical bonds of a molecule such as sugar is converted by a series of chemical reactions into energy stored in the bonds of the newly made molecules and polymers of the cells. Not with 100 percent efficiency, of course, but with an efficiency that can reach to 50 percent, greater than that of most machines. As in all machines, the rest of the energy is dissipated into heat.

This general description does not explain how the energy of the foods is used. A specific case will help to make this clear. The amino acids X, Y, Z ... as such cannot combine to make a protein because the reaction $X + Y + Z \ldots \rightarrow XYZ \ldots$ requires energy. They must first be converted to a chemical form, which may be called Xa, Ya, Za . . ., such that the reaction

$$Xa + Ya + Za + \ldots \rightarrow XYZ \ldots + 3a$$

releases energy. A reaction that releases energy can proceed, hence the protein XYZ . . . will now be synthesized. But the energy released in this synthesis must come from somewhere; it comes from energy that was expended in the reactions that converted X, Y, Z to Xa, Ya, Za.

Xa represents the *activated* form of X. This activated form is made by a series of chemical reactions, of which the crucial one involves the transfer to X of a part of a special type of molecule —a high-energy molecule, or, more correctly, a molecule with a high tendency to donate some of its atoms to other molecules.

There are many such high-energy molecules in the chemical network of the living cell, but one of them, called adenosine triphosphate, or ATP, is the central one. ATP can act as a donor of a phosphate group or of the adenosine-containing portion of its molecule. Whenever activation of molecules is required to make them participate in synthetic reactions or in any other chemical conversion, the critical step is a reaction with ATP. ATP is continually produced and continually used to activate the chemical machinery of the cells. It is made at the expense of the energy of food substances coming from outside, like the sugar in orange juice, or of those stored in the cells, like fats or sugar polymers. ATP stores this energy and releases it by reacting with other substances and making their molecules more reactive in turn. ATP is truly the energy currency of the cell.

The way ATP, the universal activator, is itself made from food substances is not always the same. The various methods that organisms use to make ATP dictate not only their chemical processes but their whole mode of life. For organisms like bacteria the way ATP is made determines whether the organism will live in soil, or milk, or in the intestine of a warm-blooded animal. For an animal, ATP production demands specific coordination between blood circulation and access to oxygen. In plants, it creates the specialization of some cells for trapping sunlight energy to feed the other cells of the organisms. Nothing contributes so much to the ecological destiny of organisms as their specialized modes of getting energy.

Many chemical events of living cells have already been discussed, including the fact that cells have chemical catalysts called enzymes, which are made of protein. The role of the enzymes in the utilization of energy can be illustrated by considering a chemical reaction such as $Xa + Ya \rightarrow XY + 2a$. This reaction will proceed because it releases energy, but, left to itself, the mixture of Xa and Ya in water will react extremely slowly. The appropriate enzyme speeds up the reaction hundreds or thousands of times. In this way the enormous amounts of chemical change that must take place in living organisms can be made to happen within a reasonable time.

The most remarkable aspect of enzymes is their specificity. Each enzyme catalyzes one specific chemical reaction involving a specific set of substances. The enzyme recognizes these substances and no others. The specificity of the recognition depends on the fine details of the surface of the enzyme protein: the substrate substances must fit precisely, atom to atom, in the appropriate niche of the "active site" of the enzyme. Then the enzyme, itself distorted by this association with its substrate, forces in turn a specific change in the shape of the substrate molecules and thereby enhances their chemical reactivity. This causes their chemical groups to gain or lose or exchange electrons, and in that way new compounds, the products of the reaction, are formed.

Although the enzyme distorts the substrate molecules in a specific way, the outcome of a reaction remains the same whether the reaction takes place rapidly with the enzyme or slowly without it. This is true because the many billions of molecules of a substance in solution are continuously bombarded by collisions with other molecules. These collisions leave the molecules more or less "energized"—that is, distorted from their normal configuration. Among those sufficiently distorted some will acquire the same configuration as the one that would be induced by the appropriate enzyme if it were present and, therefore, will undergo chemical reaction. An increase in temperature increases the number of collisions and therefore speeds up the rate of chemical reactions.

In the specificity of enzyme catalysis resides a key feature of life. Because each enzyme is made by the action of one or a few genes, a gene controls the catalysis of a specific reaction. The functioning of a cell or an organism as a chemical factory reflects the efficiency of function of the enzymes. An enzyme that can catalyze a reaction at precisely the right rate in the right kind of environment will contribute to the over-all efficiency of the organism. Not only is the active site of an enzyme critical, but also the entire structure of the enzyme molecule: the functioning of an enzyme is often modulated by the interaction of different parts of its molecule with regulatory molecules. For example, as mentioned earlier, when a cell has a high-enough level of a certain amino acid, further production of this stops because the amino acid itself combines with one of the enzymes needed to manufacture it, distorts the molecule, and inhibits its function until the level in the cell decreases again.

The whole system of chemical catalysis and its regulation is so precise that it almost suggests purpose, and indeed a special term, *teleonomy,* has been coined to denote the pseudo-purposeful functioning of biochemical mechanisms. But, as everywhere in evolution, the purposefulness is only apparent. Again, the agency at work is natural selection. A mutation in a given gene can alter the structure and functioning and regulation of the corresponding enzyme and consequently of the whole chemical network of the cell. Most mutations will to some extent upset the existing balance of chemical processes, but some may bring about an improvement of function. If the altered genes increase even to a slight degree the reproductive performance of the organism, natural selection will tend to preserve and increase them. The essential point is that in the process of evolution the genes are generally judged, not directly by their own primary function, but by the level at which their protein products function. Here the selection is precise, stringent, and deterministic, in contrast with the mutational events, which are random, unpredictable, and probabilistic. This is the critical distinction between the roles of the two agents of biological evolution—gene mutation and protein performance—a distinction emphasized by the French biochemist Jacques Monod in the title of his book *Chance and Necessity*.

To return to the way chemical energy is obtained and used, throughout the living world there are only two major processes by which it is released and made available: *fermentation* and *respiration*. When yeast uses sugar to make alcohol and carbon dioxide, that is called fermentation. In fermentation electrons are shifted around and in the process about one-twentieth of the energy of the sugar molecules is released: one-half of it goes into heat, which is why fermentation vats must be cooled; the other half is made available as ATP, two molecules of it for each molecule of sugar consumed.

Living yeast cells can use this ATP to make more cell material if all other needed foods are present. In a fruit juice there is too little other food to allow yeast to grow, but large amounts of alcohol and carbon dioxide accumulate. The ATP is wasted and that part of the energy also goes into heat. Thus an industrial fermentation such as wine making collects what to yeast cells is but a waste product—alcohol.

What happens in yeast happens in other cells too. Muscle fibers need ATP in order to contract, and they usually get it by fermenting stored sugar. In the process they make lactic acid, the same stuff that makes milk sour when bacteria grow in it and ferment milk sugar. Lactic acid must escape from muscle into the blood, which carries it to the liver for disposal. If in strong exercise lactic acid is made faster than it can be taken away, muscle fatigue and cramps develop.

Fermentation, however, is not an efficient process. When yeast or muscle have finished fermenting sugar there is plenty of chemical energy left unutilized. This is where respiration comes in. Respiration uses oxygen to oxidize many chemical substances much more completely, transferring their electrons to oxygen to make water. In this process much more ATP is produced: about 40 molecules per sugar molecule instead of only 2 produced by fermentation. Chemically the over-all process of converting, say, sugar to carbon dioxide and water is the same as burning sugar over a flame. But in living cells this process takes place at normal temperature, the electrons are carried through a series of orderly reactions before they reach oxygen, and these reactions generate a large amount of ATP. Muscles that require enormous amounts of ATP for very hard work, such as the flight muscles of insects or birds, do not ferment sugar; they use oxygen from blood to oxidize the sugar completely. Fermentation would be much too inefficient for their needs.

The word respiration is used by biochemists in a different sense from the common one. Instead of meaning the exchange of oxygen and carbon dioxide between blood and air in the lungs, respiration in the biological sense means the oxidation of a chemical compound by means of oxygen. In all oxidations there is release of energy, and the usefulness for life depends on how much of this energy is trapped in molecules of ATP and made available for chemical use. But biochemical respiration in cells of plants and animals has another characteristic: its enzymes are concentrated in those cellular particles called mitochondria, which, as mentioned earlier, are the power plants of the cell. The ATP made in the mitochondria is released into the rest of the cell for all sorts of uses. Presumably respiration with atmospheric oxygen was first "invented" in bacteria—most of them still employ it—and some of these bacteria contributed it to other cells after being swallowed by them. These cells could proceed to form complex, multicellular animals only when blood and blood circulation were developed to bring oxygen to every cell of the body.

Respiration, at any rate, must be a relatively recent invention. The early surface of the earth, long before life started, consisted mostly of metals and rocks, but it must have contained large amounts of organic matter. Experiments have shown that many organic compounds are made from simple substances by exposing them to radiations or electric discharges such as certainly occurred in the early stages of the earth. Most important, there was no free oxygen. When the first organisms evolved, therefore, fermentation must have been the only energy-yielding mechanism available to them.

As the store of fermentable carbon compounds on earth became exhausted by the spreading of early life forms, carbon dioxide accumulated in the atmosphere. Then a new way of obtaining energy developed: *photosynthesis,* the capture of sunlight and the utilization of its energy to make ATP. This ATP was used to energize a series of reactions that capture carbon dioxide and thus return carbon atoms from the atmosphere to the cycle of living organisms.

At first, photosynthesis was a prerogative of a few bacterial species. But a second development followed—probably hundreds of millions of years later: the special form of photosynthesis that occurs today in green plants, on dry land as well as in water. The peculiarity of this photosynthesis is that while it captures carbon dioxide it releases oxygen. The advent of this process generated the oxygen now present in the atmosphere and radically changed the course of life on earth. By making respiration possible, oxygen increased enormously the amount of usable energy that organisms could derive from organic foods. Thus plants, by means of photo-synthesis, capture carbon dioxide and make the organic sub-stances that serve for food to all animals. At the same time they provide the oxygen needed by animals for maximal utilization of their food.

Like the mitochondria, the intracellular particles called *chloroplasts* that carry out photosynthesis in the cells of plants are now believed to have derived from bacteria. How this happened is still in the realm of speculation. None of the bacteria now in existence qualify as ancestors of mitochondria and chloroplasts. But the over-all continuity of the biochemical processes makes it possible to reconstruct the history of the quest for energy in the evolution of life.

Fermentation must have come first. Then, by a twist on the chemistry of fermentation, some bacteria or bacterialike orga-nisms began to make organic matter—that is, cellular substance —by capturing carbon dioxide while using the energy of a variety of chemical compounds. Next, some of these organisms "learned" the new trick of using the energy of sunlight. From these there developed photosynthetic organisms that released oxygen. These in turn made respiration possible, and the conquest of the entire earth by living organisms resulted.

If photosynthesis and respiration are so efficient in capturing the sun's energy into organic substances and extracting from them a maximum of usable energy, why have other modes of obtaining energy persisted? The answer is not simple and, as usual in matters of evolution, subject to guesswork. But at least one aspect of it is relevant to the present role of biological processes: the functioning of fermentation in decay. Dead organisms or parts of organisms such as plant leaves are continuously deposited in soil or water. Here they become prey to microorganisms, which decompose them and use them as food. Ultimately all organic matter is converted again into usable substances. Some of these processes are carried out by microbes that use air, but equally important is the work of fermentative organisms working where oxygen cannot reach—in the intestines of animals, in septic tanks, in mud flats, or in unplowed soil. Fermentations, working in such inhospitable surroundings, are an essential part of the chemical cycle on which the success of life on earth depends. A similar situation exists in the human body. By means of fermentation, the cells of the muscles and of other organs, hidden away within the body and dependent for oxygen on the circulation of blood, can retain a modicum of energetic independence at least for routine work by storing a supply of fermentable sugar.

The opportunism of natural selection has preserved everything that proved consistently useful to life. Just as it is wrong to view evolution as a process in which "stronger" organisms survive and "weaker" ones perish, there is no reason to believe that natural selection, having encountered a more effective scheme to mobilize energy for the processes of life, would discard all less efficient processes. In evolution as in human affairs, surely the greatest wisdom lies in preserving a balance of mutually reinforcing, mutually complementing ways of performing a task.

# 7 ʚ Form

Out of single atoms nature fashions crystals. Out of wood or stones or bricks man fashions houses. Out of formless clay, a sculptor models his creations. Which of these processes resembles most closely the making of a cell from its molecular constituents?

When atoms in solution come together to build a crystal they are guided only by their own physical and electrical properties, which cause them to approach groups of atoms already on the surface of the growing crystal and to bind to them by electrical and other attractive forces. Thus the crystal grows: form arises directly from molecular structure. In the building of a house or the modeling of a sculpture there is another element—a plan, a program external to the raw materials. Form arises from the structure of materials by the addition of deliberate purpose.

Every cell is generated by the division of a pre-existing cell. Each cell has a definite organization, so well suited to its function that it appears to have been made with a purpose. In producing two cells out of one, new molecules are synthesized and come together to create cellular structures—chromosomes, membranes, mitochondria, and many others. The whole cell has a more or less definite shape, which is not the simple spherical shape that a formless drop of a viscous liquid like oil in water would take if left to itself. Like a house or a sculpture, a cell has a form. That form reflects the molecular structure of the substance of the cell. A cell does not take shape spontaneously, like a crystal, by the natural association of chemical groups; some of its features play a specific directive role.

Of the various types of molecules present in the cell, the many small molecules clearly do not contribute directly to form: they are present in watery solution and do not form crystals. Of the classes of large molecules, DNA has a well-defined structure of its own, the double helix. But, except at the times when it undergoes replication or when the various genes function, the structure of DNA is closed, rigid, and monotonous. Yet DNA is the backbone of chromosomes, which are assemblies of DNA, proteins, and RNA. Individual chromosomes do have fairly constant shapes and forms, easily recognizable by an expert microscopist. The over-all shape of each chromosome certainly reflects the amount of DNA it contains and probably also reflects the sequence of specific genes. In some cells the chromosomes appear under the microscope as visibly banded, and each band is characteristically recognizable and corresponds to an individual gene or group of genes. This means that the local sequences of bases in the DNA double helix determine the specific form of the chromosome, including the amounts and types of proteins that attach themselves to the DNA backbone. But of the chemical basis of these chromosomal differences biologists still know next to nothing.

The proteins are ideal substances for cellular constructions, resembling the materials of a sculptor even more than those of an architect. Each protein has a chemical and structural entity of its own. Its amino acid chains fold and curl and twist, creating a shape and surface that are uniquely specific and exquisitely appropriate for their function. The ultimate form of a protein is reached through the mutual attraction or repulsion of its chemical groups. It is as if the clay on a sculptor's bench shaped itself spontaneously into a work of art. Even when two or more chains of amino acids cooperate in building a protein molecule, they come together by themselves by the precise fit of their contact surfaces. The hemoglobin molecules of red blood cells are an example of such a precise spontaneous association.

Both the folding of protein chains and the association of folded chain into complete molecules take place without any directive "knowledge" from outside, and without need for external energy. If the chains of a protein are separated artificially, they come together again in the original relation. If a protein chain is chemically treated so that it unfolds without breaking, it slowly refolds to its pristine form. All the knowledge necessary for refolding and reassociating the protein chains must reside in the protein chains themselves—that is to say, in the sequence of amino acids dictated by the corresponding genes.

The reason all molecules of a given protein—hemoglobin for example—look exactly alike and all fold and assemble in exactly the same way is that the chemical groups on an unfolded protein chain act in obedience to the law of physics according to which objects tend to reach a state of minimal energy. This is also the law by which a body of water in response to gravity forms a surface parallel to the surface of the sea and a small amount of liquid forms a spherical drop—the shape with minimal surface. Left to themselves, the chemical groups on a protein chain, as on any other molecule, become oriented in the way that reduces the energy of the molecule to a minimum. This is not a sphere but the form that maximizes the attractions and minimizes the repulsions between chemical groups on the molecule. Thus different proteins have the most unpredictable forms, from nearly spherical to filamentous or disc-like.

The assembly of proteins to create forms of higher complexity is not restricted to the production of complex protein molecules from several folded chains. Other tasks exist for proteins, which can be accomplished by further levels of assembly. One such task is to increase the efficiency of catalytic function. Imagine a series of chemical reactions such that the products of one must serve as substrates for the next. Efficiency can evidently be raised if the catalysts are bound together in a definite order within a single complex, in the same way that the efficiency of a factory is increased by locating the machine tools in a specific order on the production line. Such production lines of enzymes do exist. A good example is that of the enzymes that build the long chain molecules of fats and oils. Several dozens of enzyme molecules are bound together in a single complex: one enzyme picks up the new small molecule that is to be added to the growing chain, another enzyme attaches it, and another group of enzymes catalyzes a series of reactions that prepare the growing chain for the next addition. Ultimately the full-length molecule comes off the complex of enzymes just as a finished automobile comes off the production line. This complex is, in fact, better than a production line: it is a production cycle, in which the object to be built is circulated again and again until it is complete and ready to operate.

Such assemblies of enzymes are not rare. One of the most remarkable of the functional associations of proteins is the assembly responsible for muscle contraction. The actual machinery of contraction has been revealed by the electron microscope. There are in the muscle cells two different kinds of protein filaments arranged next to each other. When a muscle cell receives from a nerve a signal for contraction, the two filaments, which are called actin and myosin, slide along each other and as they move they use up ATP. The two proteins are attached to the structure of the muscle cell in such a way that, when they slide, the length of the muscle cell shortens, in the same way that a portable telescope shortens when its sections slide within each other. When contraction ends, the proteins slide back and the muscle regains its original length.

In the course of evolution, cells have discovered a variety of other means to improve the efficiency of proteins. For example, efficiency can be produced by demolition. Insulin, the hormone that suppresses the symptoms of diabetes, is a case in point. When it is first produced in the living cell, the insulin molecule consists of a single chain of eighty-four amino acids, completely without hormone activity. The molecule becomes active only after a pair of enzymes cut off a stretch of thirty-three amino acids from the middle of the chain, leaving the two ends entwined together in a new configuration. As in sculpture, the final form is created by removing material as well as by adding it—by chopping away as well as by building up.

A major use of molecular assemblies in living organisms is in the construction of mechanical supports and shells. Proteins lend themselves admirably to these functions. Silk, wool, and hair are made of protein, as is also the substance named collagen that provides the material of tendons and bone. But these proteins are something special: their amino acid chains, instead of folding upon themselves into more or less rounded molecules, associate in bundles that form fibers or sheets of great tensile strength—the strength that is found in silk and wool and in the tendons that attach the muscles to the bones. In bone itself, collagen becomes impregnated with hard substances secreted by special cells to produce the solid structures that shape the body. Here, however, the final form does not merely reflect the structure of one protein: it is produced by a whole series of physiological and genetic processes that contribute to produce the ultimate form.

The proteins that form globular molecules can also contribute to generate structure and form. In muscle, the filaments of actin are linear assemblies of many small globular protein molecules all identical to one another. Assemblies of this kind provide the protein framework of many cellular structures, including the appendages that some cells use to swim around (as sperm cells do) or to generate a flow of liquid over their surface (as in the nose and throat). In many of these structures the basic units are the so-called *microtubules,* fine tubes of protein that also make up the filaments that pull the chromosomes apart at the end of mitosis. Wherever a cell needs some sort of semirigid framework it uses microtubules as subassemblies. The microtubules are produced in association with rather mysterious cellular structures called *centrioles,* which play a role in a most unexpected variety of cellular activities. For example, centrioles and microtubules are the essential components of the structures that analyze sound in the ears. The mechanical impacts of sound waves cause a bending in a clump of microtubules. This bending produces a distortion of the cellular membrane, which in turn generates the electrical signal transmitted to the nerve fibers that proceed to the acoustic centers in the brain.

A domain where artistic and geometric skills combine in the use of proteins as building materials is in the fashioning of viruses.

Each virus, as it comes out of the cell where it is produced, has a molecule of nucleic acid encapsulated by one or more shells of protein. These protein shells can take the most remarkable forms. In one of the simplest viruses, called tobacco mosaic virus from the disease it produces in the tobacco plant, there are 2130 identical protein molecules, forming a hollow tube within which the nucleic acid molecule rests. In one of the largest viruses, the cold-producing adenovirus, the shell consists of many thousand protein molecules organized in a perfect geometric crystal-like form, with twelve vertices and twenty faces—the polyhedron called an icosahedron. At the tip of each of the vertices there is a spike made of different proteins.

Scrutiny of the organization of the shells of many viruses with the electron microscope proves that their protein molecules are assembled according to well-known principles of solid geometry, the same ones employed by roof builders to construct quasi-spherical shells of maximum strength using uniform building elements. The shells of viruses bear a close resemblance to Buckminster Fuller's domes.

The perfect geometric shape of virus shells is in its way as remarkable as the symmetrical shape of a starfish or a sea urchin. But the shape of these animals and of all complex organisms is achieved through an elaborate process of development, involving cellular interactions whose complex mechanism is not yet understood. The shape of a virus is simply the outcome of the assembly of protein molecules tending, like all molecular structures, to reach a state of minimal energy. Given the proper conditions, it is possible to reconstruct a virus from its component parts in a test tube. The reconstructed viruses are just as functional, as "alive," as those produced naturally. Thus the proteins fashioned by the viral genes come together spontaneously with the viral nucleic acid to create form, and in so doing they regenerate a live organism.

If a virus can be reconstructed, it would seem reasonable that a similar process could be used to reconstruct a cell. A virus, however, is only a stripped-down organism. Because it uses for its functions the machinery of the cells in which it grows, all it needs in its free state is to have a protective shell plus a means to enter another cell. A living cell, on the contrary, is not just an assembly: it is an open system, with a flow of energy and materials and, even more important, a history. "Every cell comes from a cell" has been one of the main tenets of biological theory for the past hundred years. Could it be disproved by finding conditions under which the essential parts of a cell, coming together under purely physico-chemical forces, recreate a functional living cell?

The question is not an idle one. It amounts to asking if, in the organization of cells as it is known today, everything is dictated by the intrinsic structure of the constituent molecules—in which case reconstruction should at least in principle be feasible—or if the pattern of organization itself has somehow become autonomous and essential. In the latter case, cellular organization once lost might not be regained by simple reassembly of molecules, because in the process of taking the cell apart an essential element of information would have been lost—information provided by pre-existing cell structures and needed to "prime" the assembly of new structures as the cell grows.

A choice between these two possibilities is not easily made. Most biochemists would probably favor the first hypothesis—that the molecular elements of a cell brought together under ideal conditions could reform a living cell. But some biologists would demur, believing that certain patterns of cellular organization have, in the course of evolution, become somewhat autonomous

—that is, essential for their own perpetuation in successive cell generations. There are some grounds for these speculations, particularly in relation to the membrane stuctures that delimit cells and their compartments.

A cellular membrane must meet many requirements. It must allow entry and exit of water and other substances. It must slide and change shape as the needs of the cell require. It must accommodate the proteins that act as carriers and transporters for specific substances and must position them so that they function properly. For all these purposes a membrane must be less rigid, more fluid, than any sheet of protein would be, and yet it must be organized.

The skeleton of all membranes is composed of a class of molecules called *phospholipids*. Phospholipids are related to ordinary fats, but whereas the molecules of fats are insoluble in water, each phospholipid molecule has a "head" and a "tail," the head electrically charged, with a strong affinity for water, the tail fatlike and withdrawing from water. A minute amount of phospholipid spread on the surface of water forms an oriented monolayer, a single layer of molecules all with their heads in the water and their tails sticking out. If instead of layering phospholipid on water one shakes them together, the result is an emulsion, but not one like the emulsions of oil in water or like home-made mayonnaise, in which the oil forms real full drops. The phosopolipid emulsion consists of closed bubbles full of water. Each bubble is surrounded by two layers of molecules, one layer facing head out, the other head in, so that all the tails point toward one another and away from the water. The closed bubble is the state of minimal energy for a suspension of phosopolipid in water.

This double-layer structure is the common backbone of all biological membranes and has many remarkable properties. If a phospholipid bubble cracks it closes up again, because the rim of the hole tends to reconstitute the state of minimal energy in which all the fatty tails are sheltered from water. For that reason cells can be punctured without being killed: the membrane reseals itself. Also, the membrane can be more or less "liquid" or "solid" depending on the bends in the tail portions of the molecules. When bacteria grow at a low temperature there are more bent tails in their phospholipids, and the reverse is true at high temperatures. It is as though they were programmed to make their membranes more or less rigid—fighting heat and cold with biochemistry. At normal temperatures membranes seem to be fluid enough to allow for fairly prompt lateral mixing of their constituents and yet retain their precisely organized form and their control over exchanges between the liquids on their two sides.

In the finished membrane the phospholipids are only the supporting part. The active, operational parts are the proteins that function in transport, movement, nerve conduction, and many other processes. As might be expected, these membrane proteins have the peculiarity that their surface has greater affinity for fats than for water. This explains why they install themselves into the phospholipid membrane, but it does not explain the precision of this installation. A transport protein, for example, must take up a certain substance on the outside of the cell, turn round to deliver it to the inside of the cell, and then return to its original position. Similarly, a protein that presides over excretion must take up some substance from inside and deliver it to the outside. In other words, each functional component of a membrane must be oriented in a precise way in order to carry out its specific function. Moreover, certain groups of membrane-associated enzymes perform as organized production lines. They must be located in precise relation to one another so that the products of one reaction can flow to the next enzyme without escaping from the membrane.

This is where the question of information arises. Either these molecular assemblies within membranes are all directed exclusively by the properties of the molecules themselves, or the pre-existing membrane structure provides an indispensable framework for building new membrane. To determine which is the case, it would be necessary to succeed in reconstructing a cell in the test tube from its molecular constituents, thus proving that the information of molecular structure is indeed fully sufficient, or to demonstrate that a nongenetic change in the spatial arrangement of certain cell structures can become hereditary. That is, if a cell could be caused to make a patch of abnormal membrane, and upon returning to normal conditions the cell and its descendants continued to make patches of the abnormal type of membrane, the altered pattern would have to be considered as an element of organizational information added to the cell without change in gene-derived information.

Some suggestive observations point in the latter direction. Certain protozoa (one-celled animals) have surface structures much more complicated than other cellular membranes, with a beautifully oriented pattern of structural elements. By artificial manipulations some groups of these elements can be made to rotate and assume a new orientation. At the next cell division the rotated elements direct the assembly of new elements in their vicinity with the same abnormal orientation, and this can continue generation after generation. The genes tell the cell what materials to make, but the existing pattern on the cellular surface tells these materials how to organize themselves at each given site, acting as the guiding instruction or "primer" to extend the local pattern.

Whether this kind of informational autonomy of cellular

organization is an exceptional situation or one that exists in many cells can only be learned by experiments which at present are difficult to devise as well as to perform. It is possible that priming mechanisms like that observed in protozoa contribute to the efficiency of assembly of certain complex structures of some other cells. Yet most aspects of cellular organization are certainly the result of spontaneous interactions between the specific molecules produced by the individual genes. A survey of the assembly processes within cells—to produce enzyme molecules, complexes of enzymes, tubes, filaments, shells, and membranes—reveals no directive mechanism for creating form other than the play of the chemical groups on the molecules themselves. Like the growth of a crystal, the sculptural work of the cell is fully automatic. But, unlike crystal growth, the outcomes of the assembly process are wonderfully diverse because they reflect the enormous diversity of the molecules that take part in it. The assembled products—even the geometrical shells of viruses—resemble not crystals but works of art. The artist, natural selection, perfects molecular forms by favoring those that perform even a shade better than others. The crystals of the inorganic world stem from the workings of immutable physical forces of attraction upon a limited variety of atoms and molecules. The sculptural features of living organisms are created by the same physical forces acting on the innumerable molecular species made available by the genes.

# 8 ❧❧ Complexity

Out of proteins, lipids, and other macromolecules, evolution has modeled cells. At some point, the conquest of new living locales and the efficient exploitation of new environmental opportunities led to the emergence of organisms in which many cells cooperate. At first, presumably, two or more identical cells simply remained joined together for maximum efficiency in carrying out some vital functions. Then in these primitive organisms some cells started to differentiate. They took up specialized chemical or structural tasks and in so doing became dependent on the other cells for some of their needs. Finally this process of differentiation led to the marvelous variety of complex organisms, from the humblest seaweed to the giant forest trees, from the simplest sponge to the vertebrate animals, and ultimately, about a million years ago, to man.

The transition from unicellular to multicellular organism was not the only way in which complexity was achieved. Among the one-celled protozoa there are degrees of structural and organizational complexity unrivaled by any cells of higher plants or animals. But the one-cell organization places insuperable limits to the size of these organisms and to their biological destiny.

However the great variety of organisms that have existed on earth came about, eventually multicellular organisms such as man developed from combinations of cells and cell products. How do the myriads of cells—about ten thousand billions—that derive from a single fertilized egg come to build specifically the body of a man, with a brain, a liver, a heart, four limbs adapted to walk, run, hit, and fashion tools?

Each organ of the body has its own shape and size, determined by the activities of many genes, with only minor variations from one individual to another. If the deviation from the norm is too great, whether because of gene mutations or disease or disturbed development, the functional performance of the organ suffers, and the individual may die or be crippled by a heart defect or an underdeveloped limb. Shape and size of organs in the adult individual of each species have been perfected by natural selection. In related organisms corresponding body structures can be recognized even when they have acquired different shapes and functions. For example, the anterior limb is an arm in man and other apes, a leg in most other mammals, a wing in birds, and a fin in fishes. The same set of bones, each stupendously modified in shape and size in the course of half a billion years of evolution, serves as support for all these different structures.

The building of an organism such as man from one initial cell, the fertilized egg, obviously cannot be a simple process of growth and multiplication to give many identical cells. This only happens at the very beginning of development, when the egg, newly fertilized while descending from the ovary into the womb, gets ready to attach itself to the wall of the womb. By the time there are a few hundred cells, the growing embryo already exhibits differences from part to part: differentiation has started.

How this comes about is not known, but one thing is certain: differentiation between cells is not a change in content of genetic materials. The cells of the liver or of the brain or of muscle all have the same genes in their nuclei. Experiments have proved the truth of this statement: a frog's egg whose nucleus has been artificially replaced with the nucleus taken from a skin or liver cell gives rise to a normal individual. These experiments also confirm what was already suspected: the differentiation of cells not only does not entail loss of genes but is at least to some extent a reversible process. If a living cell is removed from its natural environment in the body and placed in a culture fluid in which it can grow and multiply, it often loses the characteristics of the organ it came from and becomes more or less similar to the undifferentiated cells from the early embryo. Therefore the differences between cells of different organs must not be in the content of the genetic material, but in its function. The groups of genes that function in liver cells differ from those in skin cells or in nerve cells. Some genes, such as those that produce substances needed for basic functions of respiration or intake of food, must be active in all cells as long as they are alive. But cellular specialization must entail the specific functioning of certain genes and the inhibition of certain others.

The key question in the problem of development relates to the nature of the mechanisms that turn on certain genes and turn off others as the cells of the early embryo generate, little by little, an enormous variety of cell types—over a period that can last many years in man, a few weeks in mice, a few days in some insects, or a few hours in certain worms. At the moment the answer is unknown. Biologists suspect that mechanisms similar if not identical to those that turn on and off the genes of bacteria in response to changes in food are also at work in the cells of plants and animals, but no clear proof has been found. In bacteria there are two general kinds of mechanisms that regulate the activity of genes: repressors or activators that can get on or off special sites on the DNA molecules and in that way block or activate the function of nearby genes; and changes in the enzyme—RNA polymerase—that fashions the RNA messages, so that certain classes of genes are prevented from participating in the process. However, the cells of man and other animals have a thousand times as much DNA as bacteria contain, and their DNA is also less free than that in bacteria: it is collected in a nucleus and distributed among many chromosomes, in which it is coated with all sort of proteins. It would be surprising if the production of such complex cells had not required systems of signals and responses in the genetic material much more complex than those found in bacteria.

Even greater mysteries surround the chemical and molecular mechanisms that produce the actual shape, consistency, and connections of each organ of the body. Yet if molecular biology is to be the interpreter of the phenomena of life in terms of the structure and chemical reactivity of specific molecules, it must be able to interpret in these terms that most amazing work of sculpture—the shaping of a human being. In man as in all other organisms the building materials are simply cells aided by some substances that cells excrete outside their bodies. How is specific form produced from these elements?

Bone is a good starting point. It is the material that gives to the body of mammals, birds, amphibians, and most other vertebrates its overall shape. (Insects, and crustacea like shrimp and lobster, have their skeleton outside their skin, made not of bone but of a hard sugar.) Bone consists essentially of collagen, a fila-

mentous protein which, in bone, becomes impregnated with insoluble calcium salts that give to bone its rigidity. Embedded in
innumerable microscopic holes within the bone are two kinds of
cells: one kind secretes more bone substance, the other kind di-
gests it away. Both kinds of cells are always at work, not only
while the bones are growing, but even in the bones of adults. The
muscles are attached to bones by means of tendons, like levers or
springs to the jointed parts of a machine. It is easy to see how
during a period of physical training the pulls and pressures of
muscles change: not only do muscles grow in size, but the bone
itself changes shape to provide the appropriate mechanical support. If the posture of a limb is changed because of a fracture, the
bone structure readjusts to suit the new pressures or loads.

Four years ago, tripping over my dog in a dark corridor, I
broke my knee cap. The surgeon decided to remove the broken tip
of the kneecap, almost half an inch high, and to join the tendon
of the leg muscle to what remained of the kneecap. The knee
healed very well, and three years later an x-ray picture showed
that the kneecap had almost regained its original shape. Under the
influence of muscular pull the bone had reshaped itself to its
normal form and was again as good as new.

The normal shaping of organs is of course not done only or
even primarily in response to external stimuli. After all, the shape
of the adult body is already visible in the newborn baby. This
shape of bones, which determines the over-all shape of the body,
is produced by a combination of an innate, gene-determined pattern and a set of external mechanical stimuli. The genetic constitution determines the over-all plan; the external stimuli play a limited but essential role in perfecting the work of nature. Even so, it
is the genetic constitution, itself the result of natural selection, that
gives to the bone cells the ability to respond in such an exquisitely
efficient way to the demands of mechanical stimuli.

Form is not always as plastic as in bone. Among the most remarkable forms created by natural selection are feathers, which are only groups of skin cells that have assumed a new specialized task. They produce large amounts of keratin, an insoluble protein in the shape of hollow tubes, which develop into the beautiful distinctive plumage of the various kinds of birds. (The hair of mammals is a less elaborate product of the same kind of cells.) In the shaping of feathers the environment plays hardly any direct role at all. Feathers develop in the pattern characteristic of the species and, once they are developed, their size and shape are not responsive to changes in function. This is because, once they have made their keratin, the feather cells die. There are no living cells within the feathers to refashion them in the way living cells refashion bone throughout the life of an organism.

So much for rigid structures like feathers and bone. But consider the shapes of the organs inside the body. The liver is shaped like a French beret, the kidneys like bean seeds, the spleen like a slipper. These organs develop inside the cavity of the body and consist mainly of masses of cells. They do not have mechanical functions, only biochemical tasks. Their shape and surface do not play any known role in these functions. On first principles a physicist might predict that these organs would be spheres—the solid form of minimum energy. But they are not. In the patterns of embryonal development prescribed by the genes there is evidently something that decides not only the size but also the shape of each organ. If a large piece of the liver is removed from an adult rat, the liver cells, which had stopped dividing when the liver reached the adult size, start dividing again, and in about a week the liver is as large as it was before, but it does not regain the same shape.

Biologists confess their ignorance of the way the shaping process is controlled by the genes. But one thing appears certain:

there is no mysterious command from within a cell that tells it to stop or start dividing. All patterning signals are interactions between cells. The genes endow cells with the ability to produce signals—chemical messages—and to respond to signals in specific ways. The signals must control not only the differential multiplication of cells in various locations to give each organ its proper size and shape but also the chemical functions of the responding cells. Signals of this kind must cause muscle cells to produce mainly the contractile proteins of muscle, feather cells to make keratin, and liver cells to produce their specialized products.

What are these signals, to which the cells of the body respond and which represent the major means of communication between cells? One intriguing type, still mysterious but practically very important, resides in the contact between cell surfaces. If living cells, taken for example from skin or kidney, are placed in a nutrient solution on a flat glass surface, they multiply until they have produced a complete single layer of cells and then they stop growing. The stoppage occurs when cells are in contact over a sufficient area of their surface. If they went on growing and piled up on top of each other the area of contact would be increased and the inhibition of growth would increase proportionately. This contact inhibition, by controlling the multiplication of cells, must play an important role in fashioning the shape of organs. Cancer cells, which have lost the normal pattern of orderly growth and produce cellular masses with no defined organic form, do not stop growing when in contact but go on multiplying and pile up into thick layers. They have lost the ability to respond to the control system exerted by cell-to-cell contact.

How this contact inhibition works, why it does not work in cancer cells, what chemicals on the cell surface transmit and receive the signals are questions to which the answers are just beginning to be found. Apparently the signals have to do with the level of certain chemicals within the cells. Indirectly, a certain level inhibits the start of a cycle of replication of DNA. The permanent cells of the adult body—the cells of the liver, muscle, or brain—make no DNA and do not divide. Cancer cells continue to make DNA and to divide under conditions in which their normal relatives would stop. DNA synthesis is a necessary antecedent of cell division: unleash DNA synthesis and cell division ensues.

Cell contacts in the living organism involve levels of communication more subtle than would appear from this description. Electron microscopists have seen that pairs of cells in contact often have their membranes joined together so tightly that certain substances can pass from one cell to the next without going through the surrounding liquid. Communication among groups of such almost-fused cells must provide special opportunities for coordination by chemical signals.

Contact signals, however, are only one class of communication devices. An organism like man needs signals that travel between the cells of distant organs. Best known among these, but still mysterious in their modes of action, are the hormones. The thyroid gland, for example, from its location in the front of the neck pours into the blood a hormone, the only product of the human body to contain iodine, which regulates the rate of chemical function in many cells of the body. This is the purpose for which people require iodine in their diet and the reason it is added to commercial salt. If the thyroid gland produces too little hormone, one feels sluggish, tends to obesity, and may even be physically and mentally retarded. If it produces too much, the individual may be thin, hyperexcitable, and suffer from insomnia. Another gland near the thyroid produces a hormone that regulates the destruction of bone in response to the body's needs for calcium. Other hormones are produced by the adrenals, the pituitary gland, the sex organs, and the pancreas.

Enough is known about the action of some hormones that they can be used as medicines, but there is still little knowledge of how any hormone acts on the cells of the body to regulate their function. Insulin, produced by the pancreas, is known to regulate the use of sugar in the body and doctors know how to treat diabetes with insulin, but little is known of how it works. Some hormones certainly affect the function of specific genes or groups of genes in the cells on which they act. But the molecular biology of this action is still in its infancy and it will be some time before biologists understand what a hormone tells a cell. There may even be hormones still to be discovered; more types of body cells may be sending out hormonal signals than are yet recognized.

Besides communication signals by cell contact and hormones, there is a third class, provided by the nervous system. Like telegraph wires, nerve fibers permeate the body. But the nerve system of communication, unlike the telegraph, is not a network interconnecting the various organs of the body with each other. Its messages travel to or from a central station—the brain and spinal cord. It is more than a communication system: it is a coordination apparatus. It does not merely transmit the messages as it receives them: it analyzes, compares, and transforms them. Even the eye sends to the brain, not a photographic image of the light patterns it receives, but a partly analyzed record of the visual field, and the brain performs even subtler analysis before it responds. Its responses are not just stereotyped reactions—simple reflexes—but subtly appropriate sets of commands that may affect many parts of the body. A localized mechanical stimulus, like a prick on the skin of a leg, may cause a reflex motion of the leg, but usually it also initiates a much more complicated series of reactions that lead to a change in posture and involve many muscles and organs in different parts of the body. These reactions may involve the participation of the cerebral cortex, in which case one is consciously aware of what is happening.

Hormones and nerves together play the key role in one of the most remarkable features of life—the maintenance of a constant internal environment in the body. This permits the body's cells to function with a relatively constant supply of food and oxygen. The essential vehicle for these materials is the blood, and its components have many properties that play a role in stabilizing its composition. But the over-all regulatory functions are controlled by hormones carried in the blood and by nerves acting directly on a variety of cells.

The molecular biology of nerves is an especially fascinating field. The nerve signal is a local electrical disturbance, which propagates very rapidly, although much less rapidly than an electric current in a copper wire. It moves along the membrane of the nerve fibers as a local change in the membrane's permeability, so that some salts can enter or escape, with a resulting change of electrical charge. The change in salt concentrations then causes the same chemical events to happen at the adjacent sites along the nerve, and this continues till the signal reaches it destination—one or the other end of the fiber. At its far end a nerve fiber is in intimate contact with one or many other cells to which or from which messages are passed. Often the arrival of the nerve signal causes a chemical substance to be released by the nerve ending and, by exciting the next cell, to serve as the transmitter of the message. To increase the precision of transmission, these nerve endings also have an enzyme that destroys whatever transmitter substance remains after the message has passed.

A baby is born with an almost completely formed nervous system, and by the time a child is two or three years old all nerve cells of the adult are present and most of them are properly connected. How do developing nerves learn to reach their appropriate destination so that the correct part of the brain or spinal cord is connected with the correct body organ? How does a nerve fiber, starting from a cell in a given position in the spinal cord, know how to reach the proper muscle in the foot? What do the cells that it passes on its way tell the nerve fiber to guide it in its search? Do they simply discourage it from stopping by exuding some chemical signal?

Above all, how does the nervous system learn? In essence, learning is a way of establishing new or more effective associations among sensations and concepts and must in some way depend on creating new or better paths for exchange of messages among the cells of the brain. Experience translates itself in an increased complexity, structural or functional, of the pattern of connections between nerve cells. When one learns the multiplication table, are more nerve connections made or is communication in existing connections made easier? And what is memory? When we recall something that we have learned, when we recognize what we have seen or heard or felt, is this recognition the tapping of certain nerve pathways, and how does it take place?

Some time ago it was claimed that "memory molecules" made of nucleic acid could be isolated from the brains of animals, for example, that after a rat had learned to run a maze, injecting an extract of its brain into another rat would make it learn to run the maze in fewer trials than were required by normal rats. Such claims have not proved valid and no serious student of the brain is surprised. Learning consists of associations among one thousand billion cells in the brain, and memory is the faculty of reactivating these associations either in the renewed presence of the same situation or in thinking—or dreaming. It is absurd to believe that each achievement, like learning to run a maze or to solve a quadratic equation, or to gauge from a frown the displeasure of a teacher or from a smile the love of a sweetheart, could be inscribed in a molecule of nucleic acid. A more reasonable assumption is that memory is the faculty of re-entering at some points a network of connections that have been created or reinforced in the brain by previous experiences and of retracing more or less faithfully the same pathway of nerve impulses. The properties of the network, its creation under genetic and experiential influences, and its accessibility in memory all remain fascinating puzzles.

An organism, of course, is not a self-contained universe. Communication is not only between parts of individual bodies but also between the world and the individual. The external world constantly impinges with innumerable stimuli. These external stimuli must be processed. If they are chemical inputs in food, or gaseous materials in the air, they are acted upon by the chemical machinery in the intestine or in the lungs. If they are mechanical or chemical or light signals, they are processed by the sense organs and the results conveyed to the brain by nerve fibers. In mouth and nose, cells that discriminate among classes of chemicals provide the organs of taste and smell. In the eye, specialized cells convert light signals into chemical signals, whose presence and arrangement is then conveyed to appropriate parts of the brain. Mechanical impacts are felt in a number of ways. In the skin, cells connected with sensory nerve fibers provide tactual sensations. In muscle, devices respond to the stretching of muscle fibers. In the ear, an elaborate apparatus receives and analyzes the vibrations of the air of which sound is made and sends a corresponding message to the brain. Within the ear, another mechanically stimulated organ continually registers changes of position with respect to the vertical stance.

These disparate devices known as the sense organs present a challenge to the molecular biologist because each is a device for converting a given stimulus—a chemical substance, a ray of light, a muscular pull, or a sound—into a transient change of molecular arrangement that can be converted into a nerve signal. Scientists are still ignorant of what goes on when a sensory cell receives a signal. Perhaps the best understood are the light-sensitive cells of the eye, which owe their exquisite sensitivity and their ability to discriminate intensity and color of light to the visual pigment— a chemical derivative of vitamin A. If a person's diet is short in vitamin A, visual pigment is not renewed and vision may suffer. But the way a chemical or mechanical signal is converted into a nerve impulse and the steps by which the impact of the outside world is converted into sensations in the living brain and, in man at least, into conscious recognition remain mysterious.

Still another set of relations between cells of the body has been discovered in connection with surgical and medical procedures. A surgeon operating on the victim of an accident decides to do a skin graft or transplant. He transfers to the appropriate place a piece of skin taken from the patient himself, generally from the thigh. He does not take skin from a different person, as he would take the blood of a donor for a blood transfusion, because he knows that a skin transplant from another person, even from the patient's brother or father or mother, would not "take." A few weeks after the operation the piece of transplanted skin would die, because the patient's own blood cells would recognize that the grafted skin is "foreign" material, attack it, and destroy it. This rejection is called an anti-graft reaction. When the graft comes from the patient's own skin, it is recognized as "self" material and is accepted. The only exception to the rejection of grafts from other individuals is in the case of identical twins, who, having the same genes, are biochemically speaking the same person.

The anti-graft reaction applies not only to skin but to most other organs. The well-known difficulties with kidney and heart transplantation stem from the fact that the transplanted organs are recognized as foreign, despite the attempts by surgeons to reduce, by all sort of treatments, the ability of the body to reject foreign substances. One exception is the cornea, the transparent front cover of the eyeball, which when transplanted does not come in contact with blood or destructive cells; that is why banks of corneas can be kept ready to be transplanted to damaged eyes just as blood banks are available for transfusion. Another exception is blood itself. It can be transfused—the equivalent of a transplant —provided the two individuals have compatible blood groups, which are relatively nonspecific classes of red blood cells. The requirements for transfusion are not so strict as for organ transplants because the transfused blood cells are not expected to live a long time as a transplanted organ is, but only to function for a few hours or days, until the patient again starts making enough blood of his own.

Apparently the body is able to recognize as "self" those proteins and other substances that it encountered during the foetal life or in the first few weeks after birth. These include not only the person's own proteins but others that may have entered from outside, either from the mother's blood or by injection. The body becomes specifically "tolerant" to these substances and the tolerance lasts for life. The explanation seems to be that the body contains a certain type of guardian cells that produce destructive substances, called *antibodies,* directed against foreign chemicals (known as *antigens*). These guardian cells are not all alike: each line of cells produces only the specific antibody against one or a few chemicals. If a chemical is present naturally or by injection in the unborn baby, the corresponding guardian cells combine with it and somehow die and are eliminated. Thus, at birth, all guardian cells against "self" substances have disappeared, but the others remain, and the baby is ready to respond to foreign substances by producing antibodies that attack them. This ability to produce antibodies is a powerful defense against bacteria and viruses, but it presents a real problem in transplantation of organs. Evidently evolution had not provided for this situation: organ transplantation is only twenty years old, while antibodies have proved useful for several hundred million years.

It is remarkable that all individuals of a species, except identical twins, should be so different that their cells are recognized as foreign. Evidently each man is unique because his set of genes is unique. Slightly different versions of a gene will produce slightly different proteins, which are often equal or nearly so in effectiveness of function. Yet the small variation will cause a protein to be considered foreign by antibodies, which recognize the fine details of its chemical surface. A few hundred molecules of a foreign protein on the surface of a cell can cause the production of enough antibodies to destroy all cells of that alien type.

The infinite nuances of heredity that the variation in relatively few genes produces are reflected in infinite nuances of individuality detectable by transplantation of organs. The chemicals that come out of a graft to stimulate the production of antibodies must be the products of the action of many genes. The chance that two individuals are identical is infinitesimal. Not only in his thoughts, his feelings, and his will, but in the chemical markings of his body each human individual is unlike any other that has ever existed.

# 9 ∾ Origins

The forms of living things have all been created by natural selection as stages in a perennial blind struggle for the perpetuation of life on earth. By the time cells and organisms made of cells had come into existence the lines were drawn. The record of fossils embedded in rocks shows that about 400 million years ago all the major groups of animals and plants now in existence were already represented by recognizable ancestors.

Was the story of life always like that? Or was there a time when the evolution of life proceeded, not by the selection of the luckier or more favored types among the descendants of existing forms, but by the addition of new forms arising from nonliving matter—by the creation of new life, not only by the expansion and differentiation of the old?

About fifteen billion years ago (cosmologists disagree somewhat about dates and mechanisms; I present here the picture favored by most) the universe started with a "big bang," an explo-

sion out of which energy and matter emerged. Whether this occurred from a collapse of a preceding universe or from a collision
of matter and anti-matter are questions that to the biologist are
irrelevancies, because life did not exist at that time. For billion
after billion of years, matter and radiation energy evolved physically and chemically. Galaxies, including the Milky Way to which
the sun and the planet earth belong, came into being, and many
of them disappeared. In the condensing clouds of matter stars
were formed—fiery balls that in turn exploded or cooled off. In the
stars the simpler atoms collided, fused, produced heavier atoms,
and emitted radiation energy, like the light of the sun that now
supports life on earth. In the gravitational field of some stars,
satellite masses of matter already cooled beyond the stage of
atomic fusion also condensed and produced planets. On at least
one planet of one star, the sun, now five or six billion years old,
physical and chemical conditions developed that made it possible
for life to begin. Whether this situation was unique, or, as cosmologists and physical chemists argue, wherever in the universe
the right physical and chemical setting exists a chemical path to
some form of life is bound to be set in motion, are questions to
which the answers may never be found.

Even by cosmic standards life on earth is not a recent happening. Half a billion years ago the oceans already teemed with an
enormous variety of plant and animal life, which was getting ready
to conquer the drying land. Photosynthesis, the machinery for
trapping the energy of sunlight, had long existed, and algae had
flourished for millennia. But earlier, what then? The primitive
earth, a mass of steaming vapors, had gradually cooled until rocks
had formed in its mantle. The cooling had to go much further
before steam became water and the oceans came into being. Soon
this water became, in the words of the twentieth-century British

biologist and writer J. B. S. Haldane, a "hot thin soup," a solution of thousands of chemical substances that could support a beginning of life. Under the physical action of heat, electrical sparks, and ultraviolet light, carbon, nitrogen, oxygen, and hydrogen had combined to produce a variety of compounds. Among them, in all probability, were also the simplest components of present proteins and nucleic acids. Today the chemist can reproduce in his test tube this *prebiotic* synthesis of the building blocks of life—amino acids and nucleic acid components—from carbon dioxide, ammonia, cyanogen, and other simple substances which, according to astronomers, may in fact exist in the atmosphere of some other planets.

If life developed only on earth (as I believe) and not on the moon, or Mars, Venus, or Jupiter it was because only earth had the right conditions: size, distance from the sun, mixture of gases in the atmosphere. Earth was fit for life, so life arose. But how? Cells as such could not have come into being directly in the hot thin soup, since a cell as it now exists is an enormously complicated object. It is unmistakably a piece of machinery perfected to do a superior job of the business of living—to provide an optimal milieu for the synthesis of the molecules of life, just as the body of man provides an optimal milieu for the cells of his organs. Evolution already had a long history before cells came into being. In the watery solution of prebiotic organic chemicals, catalysis brought about the synthesis of complex molecular species of all sorts, including proteins and nucleic acids. The nature of the primitive catalysts that acted before proteins came into existence is not known. Almost certainly clays played an important role: chains of amino acids can be produced in the test tube in the presence of certain types of clay. The biblical story of the molding of man out of clay may contain an element of truth when seen through the prism of science. Clays are remarkable substances made up of microscopically thin plates with water-seeking chemical groups on each side. These chemical groups are probably the catalytic agents—the pre-enzymes of the distant past—relatively unspecific and ineffectual, yet of tremendous use.

How many millions of times molecules arose that might have given a start to the mainstream of life and how many of them were successful are only subjects for speculation. In order for life to evolve, a form had to come into being that could direct its own replication. In the living world of today that role is restricted to nucleic acids alone. Most geneticists believe that the successful starters in the early history of life were nucleic acid molecules that could help similar molecules to assemble by a process of autocatalysis. Early amino acid polymers, possibly made with the help of clay, might then have been replaced by true proteins—that is, by "meaningful" amino acid chains made by translating a nucleic acid template. But the truly creative advance—which may in fact have happened only once—occurred when a nucleic acid molecule "learned" to direct the assembly of a protein that in turn helped the copying of the nucleic acid itself—in other terms, a nucleic acid served as a template for the assembly of an enzyme that could then help make more nucleic acid. With this development, the first powerful biological feedback mechanism had come into being. Life was on its way.

I said early in this book that the story of evolution is the record of the few lines of descent that managed to come into being among the many that might have been but most of which were not. How much more loss of potential life must have occurred in the billions of years of precellular evolution! Billions of times molecules must have been formed that might have reached the stage of self-replication but did not. Of those that did, how many were able to continue the process? Is life as it is known from the present array of living things and from the fossil record the outcome of a unique line of descent, from a single one among the molecules that learned to reproduce themselves? One primeval gene might have evolved by joining with its own copies, which then became different from their original parent (the primitive reproductive process being probably anything but accurate) and gave rise to new genes and groups of genes. Other alternatives are also conceivable: a nucleic acid capable of generating a protein that helped copy it might have provided the replicating mechanism for other, independently arisen nucleic acids. These various nucleic acids could then become associated in a more easily reproduced complex.

Whether the lineage of life on earth arose only once from the prebiotic chaos or whether multiple events converged to generate complexity will probably never be known. Either way, precellular evolution must have been extremely slow, since the chemical syntheses were without doubt exceedingly inefficient. The great speed-up must have started with the invention of the cell, which provided the efficiency of an organized chemical factory. Cells also may have evolved many times and in many places. Nevertheless, there is no reason to doubt that all organisms that are now known have stemmed from a single successful cell lineage. Some mergers may have taken place: Mitochondria and chloroplasts are believed to be remnants of bacteria that became associated with other types of cells to create unions capable of exploiting the oxygen of the air or the light of the sun. But this must have happened hundreds of millions of years after the origin of cells, when a variety of cell types already existed. At one time, students of evolution thought that bacteria resembled some primitive forms of life. But a bacterium, with its thousands of genes and enzymes and its remarkably refined adaptations to its environment, is far from primitive.

Never again, except possibly in a test-tube experiment or on some other planet, will a sample of the earliest forms of life be available. Once a successful line of descent arose, it must have spread so rapidly—that is, in one or two billion years—as to pre-empt the vital space in which any other line might evolve. If life at the molecular level were to start today, it would have no chance to flourish.

In fact, new life would have no chance to get started. The conditions that made the start possible then are not here now. As life spread, it used up the store of prebiotic organic chemicals. Then only those forms of life could survive and prosper which were able to synthesize the needed pieces anew from inorganic substances like carbon dioxide and ammonia or nitrogen gas and thus were no longer dependent on the primeval soup. The world is no longer a soup: even if new organic matter could come into being by processes like those of the early days of the earth, myriads of organisms are now present to make it their food and assimilate it into their own substance.

Thus not only was life shaped by its environment, it itself shaped that environment and in so doing it assured its own undisputed triumph. Yet that triumph was but the right to continue to evolve, life always surviving by the skin of its teeth. Time and again conditions became critical. New invention, by mutations and recombinations in the genetic materials, were the key to survival. Consider the history of the devices by which organisms obtain energy for life. Each of these—fermentation, photosynthesis, respiration—was an improvement over the preceding one. But none of them could have become established if the inadequacies of the previous device had not brought the organisms to impending starvation. Evolution works by threats, not by promises, and yet its result is the opening of greener and greener pastures.

Photosynthesis, for example, was a crucial jump in chemical evolution; it enabled cells to re-create within themselves something resembling the primeval soup of organic substances, using the same force which created that in the first place—the energy of sunlight. Thereby was life made independent of its early environment, bound only to the sun, and ready to conquer the earth.

The oxygen that plants released in using sunlight changed the atmosphere of the earth. Before photosynthesis, air consisted almost entirely of nitrogen and carbon dioxide. Now oxygen makes up 20 percent of it, screening the surface of the earth from lethal ultraviolet rays. Oxygen made it possible for living organisms to escape from the ocean and conquer dry land but did not force them to do so. Bacteria and other microbes use directly the oxygen dissolved in water. In fishes and other aquatic animals, oxygen passes from water to the blood in the gills. Mammals and birds and all other land animals get oxygen directly from the air, to which they return the carbon dioxide. Thus the cycle goes, green plants producing oxygen and using carbon dioxide, animals doing the reverse. The balance of the whole sustains life. But it is always a precarious balance. When a lake or a stream receives too much organic matter, microbes and algae increase disproportionately, using up the oxygen faster than it can diffuse in from the air, and the fish start to die. Life flourishes when balance persists, but unbalance and danger always loom—themselves the agents of natural selection.

From its first beginnings, through many successes, life on earth traveled its harried path, driven by natural selection, pushed by necessity, bereft of a goal, unobserved and ununderstood. Life had to wait for man to appear before any of its creatures asked the questions what is life, and what am I. Some men have believed, self-flatteringly, that man himself was the goal; that what evolution aimed at, what life meant, what the entire universe strove to produce was man, with his consciousness, his inquiring mind, his dissecting reason. I prefer to be more humble. Man is but one product, albeit a very special one, of a series of blind chances and harsh necessities. As with all evolution's creations, his biological fate is to make do, to survive as a species by the skin of his teeth. The skills provided by man's brain, itself perhaps the most marvelous achievement of evolution, can help him pull through, if he does not misuse them to maim and destroy himself. But he cannot escape his biological fate, just as earth and sun cannot escape their cosmological destiny to develop, flourish, age, and perish when their energy is spent. How long can man last? Ten thousand years, or a million, or ten million? Certainly much less than the few billion years of active radiation left to the sun.

Within the time span of man's existence on earth, he can hardly plan or hope to escape this planet's boundary by colonizing new worlds. The sun's other planets are not only uninhabited, but biologically speaking uninhabitable. They may permit some technologically maintained life—in a lunar laboratory for example—but they cannot support the evolution of earthly organisms. Man's evolution has occurred on earth, is marked by the earth, and cannot be exported from it. At best, the planets he has begun to explore can serve him as colonies traditionally serve the empires that conquer them—as sources of pride, outlets for the restless, and drains of emotional and other resources.

What about the farther worlds, planets of distant stars which may support life, possibly even intelligent life like ours? It is widely believed, although without strong justification, that life will have arisen and developed wherever conditions were suitable for it—possibly on millions of planets of as many stars. If this is true, will man ever know where these planets are in space, or communicate with their inhabitants? Twenty years ago, when radio astronomy was in its ambitious young years, radio telescopes were programmed to "listen" for signals from the great beyond, scanning the sky in the direction of stars that might have planets similar to earth, whose inhabitants might be broadcasting intelligible signals into the universe. Such signals were presumably expected to reflect intrinsic properties of matter or of thought— series of integers or prime numbers or formulas describing the laws of physics. No signal was detected; the program was stopped, although it may soon be reactivated. But the lure of the unknown remains. Are men, as conscious inquisitive beings, unique in the universe, or not? Whatever the answer may be, surely man can be proud that he has come far enough in biological evolution to ask such questions.

# 10 ∾ Man

Out of eons of evolution there arose man: not the fulfillment of a goal or the culmination of a striving for excellence, yet something exceptional, unique, and troublesome. For the first time in the history of the earth natural selection had fashioned a species of organisms that could think and analyze. Suddenly the roles were altered: from being the great silent creator nature became the object of scrutiny. With understanding man also acquired power. No longer did all living organisms have to submit to the limitations of their environment. A species had appeared that could learn to change its environment and fashion it to some extent according to its own conscious wishes. Not that natural selection as a determining agent of evolution had ceased to operate, even on man himself, but the criteria that fostered differential reproduction changed. Man discovered means of combatting some of the strictures posed by his surroundings and of softening the harshest impacts of natural selection on his own species. His devices for adapting his environment to his own needs differed from those of such animals as bees, birds, or beavers, which construct nests or shelters mainly through stereotyped, instinctual responses to their biological drives and to their surroundings. Man's consciousness made his relation with the environment a creative, progressive interaction. The user and inventor of tools, he devised clothing, built shelters, and extended the range of his hunting. He fought cold with fire, shortage of prey with domestication of animals, and ultimately learned to fight hunger with agriculture.

Each of these conquests required millennia for its accomplishment. From the first appearance of creatures that might be called men to the development of agriculture hundreds of thousands of years had to pass. Only in the last 10,000 years did men learn to grow abundant crops for their basic food needs.

Progress was slow, but it was true progress. Compared with the changes brought about by natural selection, the changes in human skills had a novel characteristic: they were transmitted not only biologically through the genes but also culturally by example and by word of mouth. With the invention of language cultural evolution began, not negating biological evolution but superimposing itself upon it and, by its swifter course, all but obscuring the effects of the older form of evolution.

Yet for many thousands of years man made slow progress in his technical control over nature. He invented the wheel, which made possible the first machines and some modest harnessing of the powers of water and wind. He learned to smelt and fashion copper, bronze, and finally iron. Each of these steps enabled him to occupy new niches on the earth's surface and to increase modestly his total numbers.

Then, within the last three hundred years, a new tool was

created by man's intellect: modern science, which emerged from the adoption of the experimental method and the concomitant invention of mathematical calculus. Experimentation removed the fetters of prejudice from the purview of science and technology. Calculus gave men mastery over movement and change, over matter and time in flow rather than at rest. These new tools brought about a qualitative change in mankind's power, including the harnessing of new fuels for mechanical work. No longer was man dependent on animal power, his own or that of horse or camel or buffalo, for pulling, plowing, digging, and building. His refashioning of the earth's surface could go beyond dredging ponds, building bridges, spreading the areas of cultivation—and, incidentally, destroying the primeval forests for fuel and building materials.

Science also made man aware of the limitations of his own powers. At each expansion of his powers over nature man has met new obstacles. For a long time the limitations were transient, requiring the conquest of new frontiers or the development of special techniques. Eventually, however, the conquest of the earth by man encountered barriers of a new kind, barriers that have to do with the very dimensions of the earth at his disposal. Here a new wisdom is required. The powers of man may remain wonderful tools or become destructive forces depending on how he uses them.

Nothing in human history is as remarkable or as fateful as the growth of world population—the result of modern agriculture and hygiene and medical science. Public health measures became a well-established and effective branch of science. Medicine grew from a superstition-ridden art to one based on a solid body of scientific knowledge. Today more and more medical practices are founded on the chemistry and molecular biology of the body.

Some diseases disappear or lose their role as cripplers. Immunization practices, plus the coming of antibiotic substances in the last four decades, have brought many infections under control and saved patients that would previously have been doomed to an early death. In the developed countries at least, the illnesses of late middle age, heart diseases and cancer, have replaced infections as the major killers.

With increased food, greater health and sanitation, and better medicine, the human population started growing. After centuries of relative stability the number of men and women went from an estimated 700 million in 1750 to 1 billion in 1850, 2 billion in 1930, and now almost 4 billion. If present trends continue there may be 7 billion people in the year 2000 and 15 billion or more in 2050.

The capacity of the earth may or may not be able to keep pace with this growth in terms of food, raw materials, and fuels, depending on man's wisdom and ingenuity. But humanity is clearly approaching some kind of limit. Like money at compound interest, the population increases at an accelerated rate, and earth does not grow. Even if food and heat and shelter could be supplied for the new uncountable masses, life would inevitably become enormously different and, in all probability, almost unbearable.

The main challenge of the coming decades may well be, not just to feed the billions, but to keep them from coming into existence. For the first time in evolution a species has encountered and realized the necessity of regulating its own numerical expansion. Today the technical means for limiting the size of the human population exist in a variety of birth-controlling devices, which will certainly be further perfected in the near future. Sociologists and also biologists, I among them, believe that the successful achievement of a humane regulation and limitation of human population growth is the most urgent prerequisite to attaining harmony with the world and to perpetuating an acceptable human life.

Yet the path to achieving a stable human population is fraught with difficulties and pitfalls, not biological but social and political. Human society, although warned of the impending dangers of the population explosion with its economical and ecological implications, finds within itself many forces that prevent it from turning its full attention to the impending danger.

If the population of man on earth could be stabilized, what would the biological consequences be for man? This depends on how the stabilization is achieved, whether by voluntary actions or by imposed rules and, if the latter, on the nature of the rules. The process would inevitably have some features resembling the domestication of animals or plants. The actual process of limiting the number of men and women on earth, even by voluntary means, would probably lead to some small degree of genetic selection. More ominous possibilities arise if, in devising ways to restrict its own reproduction, mankind attempted to encourage the preferential reproduction of certain individuals or groups on the basis of physical or intellectual or racial characteristics. Knowledge of human genetics and especially of the genetics of behavioral characters provides no scientific basis for *eugenics*—that is, for the selection of one group over another—or for predicting what the population derived from one group will be and how it will fare. The real dangers are social rather than biological. In human societies as they exist today, it is only too likely that the need for reproductive limitation would present, not reasons, but opportunities for new forms of persecution—of people with black or red or white skins, of Jews or Arabs or Gentiles, of Czechs or American Indians or Vietnamese. From being justified by the will of God or the laws of the jungle or the needs of the Pentagon, genocide may come to be justified by the claims of eugenics.

One might make a paradoxical case for selecting as the source of new generations those members of any persecuted group who have successfully withstood genocidal pressures. Their genetic constitution might provide gene combinations more suitable to the potentially dreadful pressures of the ages ahead. But this argument is again based on a misunderstanding. What is good in one kind of evolutionary pinch may be useless in a different one. What natural selection demands is adaptability—a range of genotypes that can do reasonably well in a range of circumstances—and plasticity—a range of genetic variation that responds promptly to changes in environment by changes in gene frequencies. The biological key to survival is multiplicity of genotypes, creative diversity, and this is even more important under conditions of controlled population size than in the wild. Domestication of selected breeds is not the way to biological success. How could the poultry industry meet new commercial or ecological requirements or new diseases of chickens if it restricted itself to pure White Leghorn flocks?

Mankind is a promiscuous species, in the sense that mobility encourages matings between members of different local populations. This practice assures the presence of a great variety of genotypes, providing the needed range of adaptiveness to new environmental situations. Selecting as progenitors of future generations the members of one or another group or type would probably not substantially reduce the over-all range of variability, but it might eliminate entirely some components of genetic variation that could prove useful or desirable in a future time.

Geneticists take a dim view of attempts to categorize groups of human beings, such as nations or races, as superior or inferior to one another, in the same way that they do not consider one form of a gene as intrinsically better than another, unless, of course, one happens to be grossly defective. The precious thing for a species in a natural setting is the maintenance of abundant genetic variability. Interfertile races, such as the great races of man, are populations that, living in different surroundings, have under the pressure of natural selection become different in gene frequencies— that is, in the relative proportions of various forms of certain genes. These populations provide the species with abundant reservoirs of genetic variability and genetic opportunities. If mankind succeeds in controlling its own population growth, it should try to do so not by converting itself into a domestic breed of selected genetic composition, but by preserving the genetic structure of a natural species such as mankind certainly is.

None of this means that in any population-control program genetic considerations should not enter. Certain genetic defects could be reasonable cause for discouraging, by genetic counseling, reproduction by individuals who are likely to produce defective children. Even here, of course, a voluntary approach is sounder than a prescriptive one.

Traditionally, human society through medicine has taken the opposite direction. Many of the triumphs of medicine have to do with correcting the consequences of genetic defects—the practice of so-called *euphenics.* In this way, of course, the affected individual is not only preserved but often enabled to produce children with the same defect. When insulin is injected into diabetics, the deficiency of the product of a single gene is corrected. When a newborn baby is taken off milk because of a genetic disease— galactosemia—that renders milk sugar toxic to the brain, compensation is being provided for the mutation of another individual gene. There is nothing irrational in trying to keep individual carriers of a genetic defect alive while endeavoring to prevent such individuals from being conceived. Once born, a child is more than a statistical sample: it is an embodiment of hope and love, of our feeling for humanity itself.

In fact, even if population controls are achieved euphenics will continue to develop. A case in point is sickle cell anemia, a genetic abnormality caused by the change of just one amino acid in the molecules of hemoglobin. When the abnormal gene is present in single copy it causes a mild respiratory deficiency; in two copies it produces a deadly illness. The reason is that the abnormal molecules tend to clump with one another and distort the cell shape. A euphenic treatment would be the administration of a drug that, entering the red blood cells, prevented the aggregation: no such treatment is yet available. Remarkably, almost 10 percent of American Negroes have the sickle cell gene. This has been traced to the fact that the sickle cell gene in single copy protects against malaria, so that people that have lived for many generations in malarial areas such as equatorial west Africa have been selected for high incidence of this gene. The malarial parasite spends part of its life within red blood cells and apparently does not like the abnormal hemoglobin.

Sickle cell anemia is a disease that should be amenable to treatment when a drug is found. Some genetic defects, however, will probably never yield to euphenic treatment. Mongolism, for example, is a congenital disease caused by a reshuffling of pieces of chromosomes in the egg. It is most frequent in children of older women: because all eggs that a woman will produce are present in her ovaries at birth, the older the egg the more chance it has had to undergo damage. What can be done about a genetic illness of this kind? One possibility, besides restraint of conception in older but still fertile women, is early diagnosis and the opportunity for abortion. A trained physician can extract and examine a sample of the clear fluid that surrounds the baby in the womb and in which float some of the embryo's skin cells. In these cells an expert microscopist can discover the abnormal chromosome. In some hospitals this test is now available to all pregnant women past thirty-five years of age. But suppose an embryo shows the chromosomal sign of mongolism, shall it be allowed to live, or shall the pregnancy be aborted?

This situation is at the borderline between birth control, usually intended as control of conception, and selective control of an existing foetus. Opinions vary, depending on ethical and religious differences, involving not only medical situations but also the question of the right of a woman to control the product of her womb. Legalized abortion will probably be an integral part of any future program of birth limitation, but resistance to abortion may well delay the adoption of a meaningful population control program. Cases such as mongolism, where abortion mercifully prevents a life of misery and a burden to family and to society, may serve to pave the way to rational future legislation.

Abortion is a negative kind of genetic intervention. It eliminates the defect by getting rid of the defective. Would it be feasible, instead, to correct genetic defects, not by supplying or altering gene products but by changing the genes themselves? This is the field of so-called genetic intervention or genetic engineering. For the moment it is only a distant possibility, but there are reasons to believe that it may soon become reality.

How genetic engineering might be done can easily be guessed. A normal gene might be introduced into the living body of an individual by injecting a purified solution of normal gene DNA. A single gene has already been isolated from the bacterium *Escherichia coli*. This was a tour de force achieved under uniquely favorable circumstances: isolation of individual genes of man will certainly be incomparably harder. But in science, once a task has been clearly defined, its accomplishment is generally only a matter of time.

Even if purified genes were available, it would not be easy to make them enter the appropriate cells to take over the function of their abnormal counterparts. DNA outside a cell is an inert substance: it does not readily enter cells on its own accord. Some better tricks than the injection of DNA would be needed, and already some such tricks can be visualized, if not put into practice. One consists in using viruses. In certain viruses part of the nucleic acid can be replaced with a piece of cellular DNA. Such gene-transporting viruses may ultimately prove to be better tools than DNA alone for correction of defective human genes because they may be able to multiply and to contribute to certain cells the genes that they carry.

The technical difficulties to be overcome before such methods become available and usable in medical practice are still enormous —selecting the proper virus, joining it with the desired gene, preventing damage to the patient by the virus itself. Treating people with live viruses may sound dangerous, but this has been practiced with smallpox vaccine for almost two centuries and more recently with polio vaccine; inoculation with a harmless live virus can provide immunity against a deadly one.

Genetic engineering might attempt to go a step further in delivering a normal gene to the individual that needs it. The problem of delivering genes into cells might be bypassed by injecting, not genes or gene-virus combinations, but cells that carry the right genes. If these cells were taken from another individual, they would probably be recognized as non-self and destroyed. But it may be feasible to take cells from a patient, grow them in the laboratory, introduce into them the desired gene or genes, and return them to the patient's body. All this sounds like science fiction. Science fiction, however, has a disconcerting tendency to come true sooner than is expected. Genetic intervention may be of some medical use within twenty or thirty years.

From correcting genetic defects in an individual a possible next stage would be correcting or replacing genes in the hereditary material that an individual transmits to his descendants. This would require efficient introduction of the right genes into the actual sex cells, the sperm or eggs. The difficulties now seem very great, but again they may vanish as biology advances. Artificial fertilization of human eggs by human sperm in the test tube is currently being perfected. These eggs, returned to the womb, would presumably develop into babies. This has not yet been accomplished and certainly should not be attempted until there is some guarantee that defective babies would not result. The time when egg and sperm meet may be a favorable occasion to manipulate their genes by correcting them or adding others according to medical needs.

Projecting the present knowledge of biology into the genetic medicine of the future leads inevitably into controversial territory. Such practices as abortion to prevent the birth of defective babies or introduction of different genes into an individual's heredity raise serious questions of moral choice. Some day it may be in human power to decide, not only who will or will not be born, but also what those born will be like, what genes they will or will not have, whom they will resemble. Another intriguing and disconcerting possibility is *cloning* of human beings by removing the nucleus from a human egg, replacing it with the nucleus from an adult cell, and implanting the egg into the womb. Such an egg would produce an individual genetically identical to the donor of the nucleus. As was mentioned earlier, cloning has been done successfully in frogs. If it were feasible in humans it could be used to produce many copies of the same individual, many identical twins, or rather multiplets. What would be the status of these clonal twins? Whether they would suffer from not being unique, or conversely draw from their biological identity with multiple siblings a new sense of human communion cannot be predicted. In general it is impossible to foresee the consequences of potential new biotechnologies because there is no way to evaluate the interactions with the social settings in which they may become available and be applied. In a society based on cooperativeness and mutual helpfulness, genetic identity might add a biological substrate to the predominant values. Unfortunately, anthropologists have found that societies approaching such ideals exist only in a few isolated populations of South Sea islanders. In a competitive, caste-ridden, power-dominated society, the ability to refashion human beings either by selection or by manipulations of eggs, sperm, and genes might become a tool to promote inequality and oppression. It might serve to create masses of obediently toiling slaves or to manufacture elites of identical rulers—in ancient Egypt, pharaohs married their sisters in order to generate successors as similar as possible to their own divine selves.

Apart from such nightmarish speculations, the philosophical implication of experiments on human heredity done for eugenic rather than medical purposes are sobering. What would a world be like in which men—even a few men—were manufactured for some alien purpose, used as means rather than ends? This is the ultimate dilemma, the same one that renders experimentation on animals acceptable and experimentation on humans not. For man is not just an animal. To men, each man is more than that because he shares with all men a conscious respect for their common humanity.

Any contemplation of the progress of science reveals that man's wisdom in regulating himself has not kept pace with his acquisition of the power to alter himself and his surroundings— through science, technology, medicine, contraception, or the genetic engineering of the future. The greater these powers, the greater are the risks as well as the opportunities. As recently as the turn of the twentieth century, with the rise of labor-saving machinery and of international associations of workers and intellectuals, and despite increasing armaments, competing territorial ambitions and colonial aberrations, the Western world expected a continuous if uneven transition to an era of peaceful progress. Two world wars and several genocides later mankind stands appalled at the capacity for mischief that has been developed.

Some writers, in discussing comparative behavior, have advanced the thesis that human aggressivity, man's inhumanity to man as seen in war and other situations, is the expression of a biological trait and is analogous to aggressive behaviors seen in many animal species. By a scientifically unwarranted jump, the proponents of this thesis have put forward in various forms the conclusion that war and criminality and racial strife are the expressions of an incorrigible biological urge. This line of reasoning has neither biological foundation nor sociological wisdom. There is no reason to believe that aggressive behaviors observed in distant species express similar genetic mechanisms. In the many thousands of generations since man's ancestry branched off from that of other primates, evolution, in developing the human brain, can have a hundred times erased the ancestral "aggression" genes if any such existed — and may even have fashioned new ones. More probably, evolution may have selected for cooperative behavior which favored communal living.

Human behavior is certainly under partial biological control, but this does not mean that it is analogous to animal behavior. Cultural and social factors play the dominant role. Aggression in human society is due much less to biological imperatives than to sociological imperialisms — that is, to the organization of society itself. Theories that ascribe human strifes to biological factors can readily be used to explain and justify racial and national conflicts on pseudoscientific grounds and to imply that such conflicts cannot be prevented by education and social decision, but only by selecting supposedly superior genotypes. Such fatalistic "biologism" has no justification in serious biology. There is no reason to doubt that conflicts in human societies have their main source in the structure of these societies and in the accompanying super-structure of beliefs, myths, and prejudices.

Irrespective of what the prime causes of conflicts in human society may be, it is not surprising that thoughtful individuals should be skeptical as to the promises of biological engineering and fearful of its potential misuse. Like the power to release energy from the atomic nucleus, the power to alter human heredity can be used for evil or for good—for destruction and debasement of human life or for its enhancement. The new powers promised by biology, if and when they become actuality, will be employed to foster whatever goals mankind sets for itself. They may be used not only to alter the physical quality of individuals but also, directly or indirectly, to fashion their intellectual, spiritual, and emotional qualities, their wishes, tastes, values—in a word, their personalities. The biologist has the responsibility of educating the public about the various options so that sensible decisions will become more likely.

To the question could man survive as man if he became a partly manufactured product of genetic intervention, the only answer available is that man is already the manufactured product of education, of upbringing, of the society that surrounds him with its constraints, promises, rewards, and deceptions. For better or for worse, with his evolving genetic heredity and his expanding cultural heredity, man blunders along, making whatever progress he can, not along a path to perfection but along the ancient biolog-ical path to survival. Yet in him there is something unique, the power of conscious thinking, that converts his passing on earth from a passive submission to the forces of nature into an adventure full of dignity and even of hope.

# 11 ∞ Mind

With the coming of man there appeared on earth a new force: the human mind. This unique instrument gave for the first time to a biological species the power to alter its relation to the environment not only by migration but by conscious manipulation of the surrounding world and ultimately perhaps even by manipulation of its own heredity. But in addition to power, and more marvelous still, the human mind generated the means for the expression of individual emotions in a collectively interpretable form—what is called art. Physical power through technology and emotional experience through art are the products of consciousness and imagination: the capacity to observe one's experiences and to categorize them in conceptual form, the gift of recalling past experiences and anticipating new ones without having to be presented with them in concrete situations.

Neither consciousness nor imagination are fully unique to man: dogs do dream, and many animals presumably carry out simple acts of abstract thought. Memory certainly plays a major role in animal activities. But the human mind is qualitatively different in its performance because it can deliberately recall past experiences and organize their traces into new patterns, either as forecasts of the future or as abstract generalizations in the search for a meaning. The human mind generates interpretations, and meanings, and values.

In its development, the critical step must have been the invention of language. Human language is an extraordinary and unique instrument. It differs in fundamental ways from the systems of communication of other animal species, such as the fully inherited mating calls of some birds or the songs that some other birds learn in their nests. Bird songs are highly stereotyped messages. No new combinations of sounds can be generated in order to communicate new or more complex pieces of information. Even the highest apes, whose brain resembles man's in many respects, cannot be taught to use words except in a most limited way. Years of language training of a chimpanzee produce results inferior to the performance of a one-year-old baby. Chimpanzees are said to learn somewhat better in sign language and may acquire a communication system comparable to that of a two- to four-year-old child.

Human language is unique in flexibility and creativity. It provides an abstract representation of objects and relationships that can be manipulated to recall, imagine, foresee, and plan. Adam, giving a name to each animal and plant in the Garden of Eden, was indeed exerting in the newly created universe the most distinctly human faculty. In providing the instrument for a symbolic knowledge of the world, language made this knowledge transmissible. No longer did an experience have to be lived in order to be known. Personal experience could be communicated to others by word of mouth to elicit wonder, transmit warnings, or impart valuable know-how, and this communication was not only horizontal, between members of a generation, but vertical, from one generation to the next. For the first time in the history of life accumulated experience could be transferred to the young by education. Side by side with biological evolution, the accumulation of gene differences, cultural evolution, the accumulation of experiences and ideas in symbolic form, had begun.

In the species *Homo sapiens* a new way of adaptation had arisen, much faster than slow-working natural selection. Intelligence could accumulate knowledge and thereby provide fitness of a new kind, permitting man to alter his environment rather than simply being selected by it. Coupled with the stupendous evolution of the human hand, that most exquisitely delicate of all tools, the development of intelligence made it possible for the species of man to spread over the earth, from the tropics to the polar regions, making the whole globe its domain.

The biological basis of the process that made the human mind what it is was the explosive development of the human brain, presumably brought about by selection for greater and greater skills. In weight, but above all in complexity, the brain of man is unique. A few million years ago, more than a hundred thousand generations, and after a much longer period of relative stability, the hominoid brain started to grow to the enormous proportions it has today, about one-fortieth the weight of the body. This growth has involved mainly those parts of the brain concerned with the higher functions of cognition and coordination—the cortex. The idea that some sort of directional process must have taken place seems inescapable. Biologically speaking, this means that once certain mutations started to produce a more powerful brain system, this system proved so valuable for differential reproduction that any new gene combinations that perfected it further were powerfully favored. One might almost say that in the recent evolution of man practically everything else was neglected in favor of increased brain power. Man lost the protective fur of the apes, their early sexual maturity, and many other adaptations useful to lower mammals. In exchange he won the brain and with it the faculty of language, speech, thought, and consciousness.

The central role of speech and language in the development of thought-power and in the success of man as a species suggests that a major part of the evolution of the human brain from that of man's ape-like ancestors must have been a continuous perfecting of the speech centers, which are located on the left side of the brain. The location of some of these centers is known from the symptoms observed in individuals who either from strokes or from accidents have received localized brain injuries. These injuries result in various kinds of speech trouble, the so-called aphasias, their nature depending on the exact site of the damage.

That evolution, focussed on perfecting one specialized faculty such as language ability, can produce an enormous development of the corresponding part of the brain is not an unfounded speculation. There are other examples in the animal kingdom. Electric fishes, which depend on perception of an electrical field for information about the world around them, have a spectacular enlargement of the portions of their brain concerned with emitting, receiving, and analyzing the electrical signals. Other instances could be cited. Evolution does not, of course, "aim" at perfecting a certain function; rather, its workings are channeled in a certain direction because of the level of performance already at hand. The existing mechanism acts as a multiplier for the results of added genetic components, in the same way that an established, effective economic structure increases the potential of any added capital.

The developed capacity for symbolic language—human language—may therefore he called the biological basis for human culture. But here arises the problem of what biology does in fact contribute to human language. Even more than a song-learning bird, man is a language-learning animal. Bereft of human companionship, a child neither learns nor develops a natural language of his own. He may at most imitate the noises of animals or the winds or the ocean. Why then do biologists believe that the structure of language is not fully learned by experience but is in part at least imbedded in the network of connections of the human brain?

Modern linguistic analysis has revealed that all human languages share common fundamental patterns of grammatical relations. The most primitive dialects of African tribes or Australian aborigines have in common with French or English a basic formal structure, suggesting that this structure is the expression of a common organization of the human brain. Moreover, given some elements of language by contact with other human beings, a child in the course of his development "invents" new combinations, new assemblies of functional parts of the established language. Thus the language that each individual develops is partly learned and partly an expression of the structure of his own brain, and it has general validity in communicating with others because the language structure of the brain is to a large extent common to all men.

Language is an assembly of verbal structures representing objects and relationships. From thinking of language as a dual entity consisting of a genetically determined component inscribed in the structure of the brain and a learned component derived from experience it is an easy step to a more general conception of the human mind. As men conceptualize in thought or speech the world around and within them, the mind analyzes these perceptions in terms of certain schemes of properties. From Aristotle to

Descartes to Immanuel Kant, philosophers have long perceived these analytical categories, the elementary structures of mental analysis, as something given rather than learned. To the biologist it makes eminent sense to think that, as for language structures, so also for logical structures there exist in the brain network some patterns of connections that are genetically determined and have been selected by evolution as effective instruments for dealing with the events of life. Insofar as logical structures are dealt with in terms of language, this cerebral substrate of logic would be an integral component of the language structure. Perfecting of these cerebral structures must have depended on their becoming progressively more useful in terms of reproductive success. For language this must have meant becoming a better instrument in formulation and communication of meaning through a usable grammar and syntax. For logical structures in general, selection must have been for more effective "thinking."

To make this picture of the biological component of the mind more precise requires closer consideration of the structure and development of the nervous system. The network consists of nerve cells, each of which sends out a nerve fiber more or less branched. Signals —a taste or a pain or a command to move a finger—travel along fibers as groups of electrical impulses, which can be recorded with fine electrodes. The nature of the information being transmitted is represented by the frequency and the spacing of impulses in each fiber and by the number and identity of the fibers involved. Contacts between nerve cells, required to create a network, take place where a branch of a nerve fiber terminates on the body of another nerve cell. The contact is called a synapse. The signal passed on at the synapse may excite the receiving cell to "fire" and to send out in turn a signal through its own fiber to a new set of receiving cells, or, if it is an inhibitory signal, may prevent the receiving cell from firing in response to excitatory signals from other cells.

Even at this elementary level the degree of precision that such a network can generate becomes apparent. A given cell may be wired to fire only when stimulated by a given set of N cells, provided it is not inhibited by some other set of M cells. One of the best analyzed systems, the visual apparatus of a mammal such as the cat, gives ample proof of the correctness of this scheme. As already mentioned, between eye and brain there is a hierarchy of levels. Each level analyzes the over-all properties of the light signal, telling the next level more and more details of the amount of contrast between light and dark areas on the eye's retina, of the spatial relation between lighter and darker areas, and of the motion of boundaries between them. Some cells in the brain cortex concerned with vision may actually receive and process information derived from the entire visual field and survey in exquisite detail the total features of the visual experience, in addition to receiving information from parts of the brain where other sensations are processed. The firing of such all-knowing cells, if they in fact exist, might send signals to all parts of the brain concerned with appropriate action, or with storing the memory of the visual experience, or with the integration of visual experience with other sensory inputs.

What is important here with regard to the biological basis of the human mind is the extent to which the network of the brain, compared to which the most elaborate computer is but a kindergarten toy, is dictated by heredity. The very facts that the over-all structure of the brain is the same in all normal individuals, that the nerves that reach various organs are always the same, and that the basic pattern of the nervous system is laid out well before birth point to the role of heredity in network formation. Experiments have amply confirmed this role. Nerves severed from their terminal organs regenerate and again reach the same organs to reestablish more or less precisely the original connections and function. The nature of the stimulus that causes the specific attraction of given nerve fiber to a given organ and even a given cell within an organ remains unknown. Perhaps each site has an individual, recognizable chemical make-up, but whether that consists of a difference of concentration of one or more chemical substances and how such a chemical gradient can be recognized are questions that await future research.

If the peripheral nervous network is in the main genetically determined, the central network within the brain certainly is. Most connections develop before birth, long before any external stimuli are available. These inherited patterns of connections conceivably include the networks that underlie the structure of language and the structure of reasoning in the same way that they include the network underlying the analysis of visual or auditory experiences.

However, to state that the basic network for language or for logical thinking is inscribed in the structure of the human brain does not mean that the brain has encoded in it the English language or some "natural" language. Nor does the ability to learn to play chess mean that the rules of the game exist in the nervous network. What is present in the brain is a set of interlocked systems of connections that can be programmed by experience, one to become an effective instrument for spoken or written language, another an instrument for mathematical logic, a scheme for chess playing, and so forth.

What then is the role of experience in the learning process? What is added to the inborn network of linguistic and logical faculties by learning English, or Swahili, or chess, or calculus?

More generally, is learning the establishing of new synapses among nerve fibers and cells that were not in contact before, or a facilitation of the passage of impulses at synapses that already existed, or a blockage of inhibitory connections? Despite all the progress in the analysis of brain function these crucial questions are still unanswered and will probably remain so until the simplest networks possible, those established among nerve cells growing in a culture dish, have been successfully analyzed.

An essential part of learning is memory, the ability to retrace associations established by previous experience. Such retracing need not be conscious. In the classical experiments on conditioned reflexes, the simultaneous presentation to a dog of a relevant stimulus such as food and an irrelevant one such as a noise established a connection between noise and salivation, so that later the noise alone caused saliva to flow. In conscious recall, a complex pathway of associations is followed in a more or less precise way to produce rational thinking. The thinking mind must be able to select the series of connections needed for the purpose at hand and to inhibit the by-paths that can lead it astray. This is the willed element of purposeful thinking, as distinct from the passive, purposeless pathways of thought in dreaming or daydreaming. Yet even dream thinking has a certain regularity, presumably because the sequences of cells that are brought into activity represent paths of the brain network that have been established by thought activity during wakefulness. On the other hand, even in the most concentrated active thinking, the selective precision of the main line of thought is never complete: no matter how effective the control systems may be, one always remains aware during conscious thinking of the many by-ways to which a line of thought could give rise. In fact, it is this very richness of potential branchings that gives thought processes the wealth of implications that makes them creative—the suggestiveness from which poetry draws its magic, tapping the hidden connections of language to recreate and remodel the multifaceted form of experience.

Language has given man culture, and culture has evolved. The way cultural evolution superimposed itself on biological evolution has already been described. Biological evolution is blind and Darwinian, consisting only of the selection of genetic structures by differential reproduction. Cultural selection is based on the transmission and teaching of knowledge experientially gained. For this reason, cultural evolution is extraordinarily rapid. It improves the fitness of man for his environment, not by altering the genetic make-up of mankind but by altering the environment and making it more fit for man. Man's culture also exerts powerful genetic influences on other species, either by ruthless eradication or purposeful domestication.

Yet, for all his cultural achievements, man does not escape biological evolution; he only modifies its effects on himself. Whenever, for any reasons, conditions develop that are not optimal for human life—whether because of epidemics, famines, poverty of any sort, or polluted or overcrowded environments—selection of certain human genotypes and therefore of certain human genes will always occur. Mutations continue to happen in man as in all other organisms, offering new dangers and new opportunities. Man escapes biological selection only insofar as he learns to repair or prevent the consequences of environmental stresses and genetic defects. His ability to withstand the inevitable stresses depends not only on his cultural skills but also on the availability within the total genetic make-up of the species of sufficient hereditary variability to provide the necessary adaptiveness to a variety of external conditions.

At the collective level, consciousness based on language generates culture and its evolution. At the individual level, consciousness generates in man characteristics of behavior different from those of other animals. Take for example the mating behavior in pair-mating birds like doves. Here a remarkable sequence of events takes place: courtship with elaborate ceremonies leads to mating, nest making, egg laying, and sitting activities, the whole presenting to the observer the appearance of a mixture of romance and domesticity. But experimental analysis of these chains of behavior reveals that, despite superficial appearances, the whole sequence is activated by a series of hormonal responses to visual stimuli; each phase of the sexual relationship, if consummated, triggers the release of the hormone that calls forth the next step. The charming romance is fully stereotyped, entirely inscribed in the genetic make-up of the species.

In man, however, consciousness generates the awareness of choice. In sex, as in all other activities except those of early infancy, man chooses; in fact, he is forced to make choices because he is aware of alternatives even when they are not physically present. Human behavior is conscious behavior and by virtue of that fact man is more than another animal. As was stated by the British philosopher Jacob Bronowski in *The Identity of Man* (1965), each man is a "self" not only because of his biochemical markers of individuality but also because he is the repository of a unique set of experiences and is thereby endowed with a unique set of choices. Because of consciousness and individual experience, even identical twins are selves, in spite of being biologically identical or nearly so. Thus conceptual knowledge generates consciousness, which in the human collective gives rise to culture and in the human individual produces a new kind of uniqueness besides the uniqueness of gene combinations. Conscious choice and conscious restraint replace the reflex, automatic behavior of other animals.

Human behavior also includes automatic aspects, not only in the functioning of internal organs but in responses to situations that evoke immediate emotions like fear or aggression, but by and large the behavior of socialized human beings is subject to the influence of consciousness.

Does this also mean that conscious behavior is free—that is, does consciousness of choice imply freedom of choice between available alternatives, or is freedom of choice an illusion, the individual's actions being rigidly determined by his past experiences plus the biological structure of the nervous system? Nothing in biology provides an answer to this central question of philosophy. Yet nothing that is known in biology is incompatible with the idea of free will regarded as the opportunity to direct one's actions by the determinant act of electing one out of a set of choices open in a given circumstance. Nervous pathways must exist in the brain which in response to imaginative representation of the possible consequences of alternative choices dictate through the play of excitation and inhibition of the appropriate nerve cells the performance of one given set of actions.

The question, however, is not thereby resolved but only displaced. The brain may permit an individual human being to act consciously in the light of a rational expectation of certain consequences. But how are the desired consequences chosen? In other terms, what role does the brain play in a value system? Is ethics a set of mechanical rules of choice inscribed in the brain by heredity plus experience, channeling choices into predetermined paths despite an illusion of freedom? Or is ethics the expression of a function of the mind by which a human being, before acting, explores in the network of his brain the pattern of wishes, hopes, fears, abnegation, and other emotions and beliefs derived from the total experience of the self? If so, then every action could indeed be free, at least within the range of choices provided by the brain network. In fact, each action could be creative of new ethical, choice-generating pathways within the network. In this view a free action does not commit the individual to future repetition, as in a conditioning experiment, but enriches the range of choices available in the future, not only by repetition but by comparison and combination. The gratuitous act of the existentialist hero—murder or suicide or escape—is not necessary for the assertion of individual freedom. Conscious activity is the assertion of freedom because thereby does the self grow.

Here I approach the central dilemma of life science. The essence of biology is evolution, and the essence of evolution is the absence of motive and of purpose. According to Monod's formulation, chance and necessity are the two faces of biological progression—chance errors in the genetic materials and chance changes in environmental conditions; necessary rigid standards of performance of organisms within their environment. But with man purpose and motive enter the picture, and, according to the view that Bronowski stressed and with which I concur, motive and purpose —that is, values and will—are inevitable expressions of man's brain. Their specific content is the product of experience, in the same way that the specific languages that men use are products of their experience. But values, will, and language are constrained by the structure of the human brain. In constructing this marvelous organ evolution has dialectically surpassed and denied its own past history. Biologically the human brain is probably still evolving. But it is questionable whether the effects of any further biological evolution of the brain will be noticeable beneath the continuing evolution of human culture. Only if man's culture, utterly misapplied, should create extreme stress situations such as excessive overpopulation or intolerable pollution or famine, could biological selection, whether for brain power or for some other genetic traits, again become paramount in human history.

Let me come back to the question of values. I have suggested that the choices that men make, the values they adopt in the course of their activity, may be partly based in the network of brain connections, inherited or learned. Does this mean that value systems are in part at least biologically determined? My answer would be yes. In the thousands of generations during which man's evolution fashioned the brain patterns of language and consciousness and imagination as the biological substrates for intellectual activity, it also fashioned the structure of human society based on language and verbal communication and conceptual abstraction. Some fundamental features of the structure of human society are certainly dictated by biological heredity, just as the structures of animal societies are inscribed to a much higher degree in their genetic endowment. Values are frameworks for making choices between alternative opportunities. If the pattern of recognizable choices is limited by the pattern of nerve connections by which they are conceived in the mind, then also values, the criteria for decisions between choices, must be in part circumscribed by the pattern of connections through which the potential consequences of actions are interpreted.

There is no conflict between the claim to free will—the faculty of making deliberate choices between alternative courses as presented in the mind by a biologically limited set of connections—and the assertion that the values that underlie the choices are themselves limited by the biological properties of the brain network. Every aspect of the conscious activities of men presents the inescapable dualism of the mind, an instrument created by biological evolution and perfected in each individual through learning, experience, and social intercourse.

As most human beings painfully realize in their own life experience, the dualism of the mind places on men, alone among all living beings, an almost unbearable burden: the consciousness of one's individual transience, the awareness of inevitable death, what existentialist philosophers have called the "absurd" of the human condition. Yet most men do not despair. They do not rush into suicide to escape the anguish of life's absurdity. In the words that Voltaire in *Candide* (1759) put in the mouth of an old woman, raped, mutilated, a hundred times defiled: "*Je voulais cent fois me tuer, mais j'aimais encore la vie*" [A hundred times I wanted to kill myself, but I still loved life].

Humankind is justified, I believe, in suspecting that once again blind evolution has operated with subtle wisdom. While fashioning consciousness and exposing man to the ultimate terror, it may by natural selection have also brought forth in the human mind some protective compensatory features. Human evolution may have imprinted into man's brain an intrinsic program that opens to him the innermost sources of optimism—art, and joy, and hope, confidence in the powers of the mind, concern for his fellow men, and pride in the pursuit of the unique human adventure.

*Glossary*
*Suggestions for Further Reading*
*Index*

# Glossary

adapter RNA   The RNA molecules that serve to position the amino acids in their right places within proteins (see *ribonucleic acid*).

adenosine triphosphate (ATP)   The energy currency of all cells. By donating some of its atoms to other substances it allows them to participate in chemical reactions.

amino acid   A substance that, by making polymeric chains with other amino acids, gives rise to proteins.

anticodon   A group of three nucleotides on the adapter RNA which, by pairing with the corresponding codon on messenger RNA, positions the attached amino acid at its appropriate site in a growing protein chain.

antibody   A blood serum protein that is made in response to the presence of a given foreign substance (antigen) and combines specifically with that substance.

antigen   Any foreign substance capable of stimulating the production of a specific antibody.

chloroplast   The intracellular structure containing the chemical machinery for photosynthesis in plant cells.

chromosomes   Filamentous structures present in the cell nucleus and containing the genetic material.

cloning   The isolation of pure lines of cells from one single cell; also, the process of producing identical organisms by introducing into a series of egg cells nuclei derived from one single individual.

codon   A group of three nucleotides that represents a given amino acid in RNA or DNA (see *ribonucleic acid; deoxyribonucleic acid*).

convergence   The process by which similar organs are generated in distant groups of organisms by different genetic mechanisms.

deoxyribonucleic acid (DNA)   The macromolecular substance that constitutes the materials of the genes (except in some viruses); a polymer of nucleotides.

diploid   A cell or organism with two sets of chromosomes and therefore two sets of genes.

dominant   That form of a gene which, when present with another form of the same gene, masks the activity of the latter (see *recessive*).

entropy law   The second law of thermodynamics, which states that the only changes (for example, chemical reactions or heat transfers) that can occur spontaneously in an isolated system are those that increase the entropy—that is, the disorder of the system.

enzymes   Protein catalysts that modify specific reactions in living cells.

eugenics   The improvement of the genetic constitution of organisms; used especially with reference to man.

euphenics   The improvement of health by correction of the consequences of genetic defects.

fermentation   An enzymatic interconversion of organic substances that yields energy in usable form.

gene   A portion of genetic material, usually DNA, that determines the structure of a protein chain; also, a genetic element whose mutations affect some recognizable trait of an organism.

gene pool   The range of genes available in the individuals of an interbreeding population, usually a species.

genetic code   The rules of concordance between codons in nucleic acids and amino acids in proteins.

genetic drift   A change in frequency of certain genes as a result of their chance-determined presence, absence, or frequency in very small populations of a species.

genotype   The complex of all the genes of an organism.

germ cells   The cells that in the sexual process fuse together to generate a new organism; generally referred to as sperm cells and egg cells.

hemoglobin   The red protein of red blood cells, consisting of four chains of amino acids (two different pairs, each of which has two identical chains).

hydrogen bond   A weak bond formed by a hydrogen atom shared by two other atoms (O-H•••O; N-H•••O); important in determining the structure of proteins and nucleic acids.

macromolecule   An organic molecule with molecular weight over 1000 or 2000; usually a polymeric molecule.

meiosis   The process by which the germ cells are produced in the sex organs, each receiving only one set of chromosomes.

messenger RNA   RNA molecules that are transcribed off the DNA of the genes and serve to direct the synthesis of proteins.

metabolism   The entire process of chemical interconversions by which food substances give rise to building blocks for cellular macromolecules and other substances are broken down for elimination.

microtubules   Tubular protein structures, visible in the electron microscope, that play a role in many cellular processes.

mitochondrion   A structure which in oxygen-using cells produces ATP from a variety of foodstuffs.

mitosis   The process of cell division; specifically, the mechanism which partitions the chromosomes of a cell equally to the two daughter cells.

monomers   The unit elements of which polymeric molecules are made.

mutation   Any change in the genetic material; more restrictedly, any change in the structure of a gene.

nucleic acid   A polymeric substance consisting of nucleotide chains, in which adjacent nucleotides are held together by bonds between the sugar and the phosphate groups.

nucleotide   A substance consisting of a nucleic-acid base, a sugar, and a phosphate group. The sugar is ribose in RNA, deoxyribose in DNA. The bases are adenine, guanine, cytosine, and uracil in RNA; thymine instead of uracil in DNA.

phenotype   The set of gene-determined characteristics of an individual organism.

phospholipids   Substances that form the backbone of cellular membranes, one end of their molecules having affinity for water and the other end being water-insoluble.

photosynthesis   The process by which plants and certain bacteria synthesize organic matter from carbon dioxide using the energy of light.

polymer   A substance made up of chains of small molecules (monomers) in specific order and linkage.

prebiotic   Existing before life appeared on earth.

proteins   Polymers of amino acids, which constitute about 50 percent of the non-water substance of all cells; the substance of enzymes and many other cell constituents.

recessive   A form of a gene which, when present together with a different form of the same gene, has its activity masked (see *dominant*).

repressor   A protein that, by attaching itself reversibly to or near a gene, prevents its functioning.

ribonucleic acid (RNA)  Nucleic acid made as transcript from DNA genes and serving mainly to direct the synthesis of proteins; also, the genetic material of certain viruses.

species  The entire group of organisms, similar but not identical to one another, that form a potentially interbreeding population.

teleonomy  The apparent purposefulness of biological adaptation resulting from the effective functioning of natural selection.

template  A mold on which a new form is cast. Biologically, a polymeric molecule whose sequence of monomers serves to direct the synthesis of some other sequence. DNA and RNA are template molecules.

transcription  The process by which the sequence of nucleotides in DNA dictates the sequence of nucleotides in RNA; rarely the reverse.

translation  Biologically, the process by which RNA serves as template to order the amino acids in protein.

virus  An organism whose genetic material becomes encapsulated into a protein shell and can only function and reproduce when it reenters a living host cell; generally the cause of some disease of the host organism.

# Suggestions for Further Reading

Cairns, John, Stent, Gunther S., and Watson, James D. (eds.) *Phage and the Origins of Molecular Biology.* Cold Spring Harbor, N.Y.: Cold Spring Harbor Laboratory, 1966. This collection of original papers written in honor of Max Delbrück includes revealing historical recounts of many important discoveries in molecular biology.

Dobzhansky, Theodosius. *Mankind Evolving.* New York: Bantam Books (paperback), 1970. A great geneticist looks at man as a unique species of animal. "The most interesting scientific treatise that has ever been written on the nature of man"—George Gaylord Simpson.

Dubos, René. *So Human an Animal.* New York: Charles Scribner's Sons, 1968. An inspired discussion of the interplay of the biological and spiritual aspects of man, solidly grounded in the ideas of modern biological science.

Katz, Bernard. *Nerve, Muscle, and Synapse.* New York: McGraw-Hill (paperback), 1966. A clear, short, and relatively simple exposition of the structure and workings of nerves and brain.

Lehninger, Albert. *Biochemistry.* New York: Worth, 1970. Unquestionably the best textbook on the subject; for those who plan to go on to a serious study of fundamental biology. Not to be read for easy pleasure, yet superbly written, crystal clear, and almost free of mistakes. Junior year college level.

Lerner, Michael. *Heredity, Evolution, and Society.* San Francisco: Freeman, 1968. A unique kind of textbook, which makes genetics easy by a close and thoughtful reference to human concerns. First year college level.

Monod, Jacques. *Chance and Necessity.* New York: Knopf, 1971. A controversial essay on the natural philosophy of modern biology by one of the great masters of molecular biology. The overview of the structure of biology leads the author to an existential proclamation of an ethics of knowledge.

Watson, James D. *The Double Helix.* New York: Atheneum, 1968. The controversial history of the discovery of the structure of DNA, and a delightfully candid human story.

———. *Molecular Biology of the Gene.* 2nd ed. New York: Benjamin (paperback), 1970. A young "classic," with a superb presentation not only of the basic facts of molecular biology but also of their applications to cancer, immunity, and development. Undergraduate level.

# Index

## About the Author

S. E. Luria, Institute Professor, Sedgwick Professor of Biology, and Director of the Center for Cancer Research at the Massachusetts Institute of Technology, Cambridge, Massachusetts, shared the 1969 Nobel Prize in Physiology. He and two other American scientists were awarded the prize "for their discoveries concerning the replication mechanism and the genetic structure of viruses."